ELEMENTS OF TRIGONOMETRY

Elements of Trigonometry

Dr Maria Tattaris

ARCLER
PRESS

www.arclerpress.com

Elements of Trigonometry

Dr Maria Tattaris

Arcler Press
2010 Winston Park Drive,
2nd Floor
Oakville, ON L6H 5R7
Canada
www.arclerpress.com
Tel: 001-289-291-7705

001-905-616-2116

Fax: 001-289-291-7601

Email: orders@arclereducation.com

ISBN: 978-1-77361-317-8 (Hardcover)

About the Author

Maria Tattaris has a background in Mathematics and a PhD in remote sensing within the area of atmospheric sciences, awarded in 2012. Her interests include field spectroscopy, image processing, spectral analysis, field data collection, and investigating new data analysis methods for remote sensing techniques, all within the areas of environmental, plant, atmospheric and related sciences.

Contents

Preface

Trigonometry, the branch of mathematics that describes the relationship between the angles and lengths of a triangle, was originally developed thousands of years ago. It may seem at times that trigonometry is never used outside the class room. However, you may be surprised to learn how often applications of trigonometry are encountered in the real world.

Trigonometry is applied in the real world in many areas, such as architecture, engineering, astronomy, geography, physical and social sciences, surveying and navigation, highlighting the importance of the subject from colonial times to the modern world.

In the classroom, many students fail to see how simple trigonometric ideas, such as measuring angles of a triangle, can be applied to real life situations. Moreover, there is a lack of knowledge as to where these ideas come from. Rather than incorporating trigonometric concepts with its predecessors, algebra and geometry, it is often introduced as a separate unit in the mathematical world of knowledge. As a result of having no connection to previously acquired mathematical awareness, students are bemused by the particulars and use of the, not only in mathematics, but also in everyday life.

Perhaps by providing a concise insight into where these concepts come from, how they were discovered, and how their uses in the past relate to how they are presently used and taught, may offer students the understanding they need to put these concepts into use. With this improved insight, one would be able to see the value of studying trigonometry as an element of mathematics, instead of an isolated unit from the subject.

This book aims to help readers understand the wonderful world of Trigonometry, and to enlighten them on how this subject's worth in real world problems.

Author

CHAPTER 1

HISTORY AND FUNDAMENTALS OF TRIGONOMETRY

INTRODUCTION

This chapter is divided into three sections. In first section, the history of trigonometry is presented, the second section contains some real-life examples of trigonometry and last section explains reasons why trigonometry is important in our life.

Trigonometry was discovered from the ancient practice of mathematical astronomy, particularly the modeling of the motions of celestial objects. In order to be able to apply geometric models to predict the position of planets, Hipparchus and later Ptolemy needed mathematical tools that convert arc lengths to distance measurements. Much of this work took place on the sphere rather than the plane, with Menelaus's Theorem as the fundamental tool. In India, trigonometry took on a more computational and less geometric form than its Greek inheritance, but was used for similar purposes.

Interpolation and approximation allowed them to solve trigonometric problems that were otherwise unsolvable. Trigonometry was used in Islam for astronomical timekeeping, as well as for ritual purposes, such as determining the direction of Mecca and the beginning of the lunar month. The West was originally dominated by influences passing through Muslim Spain.

DEFINITION OF TRIGONOMETRY

The term trigonometry comes from the Greek word trigonon, meaning triangle, and the Greek word metria, meaning measurement. As the name implies, trigonometry ultimately developed from the study of right triangles by applying the relationship between the measure of sides and angles.

However, the word trigonometry did not exist upon the birth of the subject, but was later introduced by the German mathematician Baratholomaeus Pitiscus in the title of his work, Trigonmetria sive de solution triangularm tractatus brevis et perspicious…, published in 1595.

APPLICATIONS AND USES OF TRIGONOMETRY

Trigonometry, the branch of mathematics that describes the relationship between the angles and lengths of triangle, was originally developed thousands of years ago. It can seem like trigonometry is never used outside class room, but you may be surprised to learn how often applications of trigonometry are encountered in the real world.

- Architecture
- Engineering
- Astronomy
- Geography
- Physical and social Sciences
- Electricity
- Industry

Research Literature on Trigonometry and Related Subject

Trigonometry is an important school subject, not only in mathematics but also in other disciplines. It has many important applications in engineering, astronomy, physics, architecture, and so on. In terms of mathematics, it is one of the fundamental topics in the transition to advanced mathematics and its applications. A firm understanding of trigonometric functions is required in calculus and analysis.

Hence, trigonometry has an important place in the mathematics curriculum in many countries, even though its meaning may change from country to country at secondary school level.

Trigonometry has a complex nature. It combines different algebraic, geometric, and graphical concepts and procedures. For instance, trigonometric functions are different from other forms of functions in that they cannot be computed directly by carrying out certain arithmetic calculations revealed by an algebraic formula. Evaluating trigonometric functions requires both geometric reasoning and algebraic reasoning. This complex nature of the topic makes it challenging for students to understand it conceptually.

Research on trigonometry is sparse, and there appears to be just a few researchers investigating students' understanding relating to different instructional approaches, even though trigonometry has been often reported as a difficult subject for students. This section presents a literature background for the present study. Firstly, studies are reviewed in terms of their understanding of trigonometry.

This is followed by a review of studies examining the use of technology in the teaching of trigonometry. This is carried out due to the limited literature, and the interrelated nature of trigonometric functions with other mathematical subjects.

Understanding of Trigonometry

Despite being limited, research shows that students have a fragmented and incomplete understanding of trigonometric functions. In one of a few studies which investigated students' understanding of trigonometric functions, Weber (2005) examined college students' understanding of trigonometric functions with an experimental design based on the notion of procept. He implemented a lesson design based on predominantly hands-on activities, such as students' construction of the unit circle, and drawing angles and related line segments corresponding to their trigonometric values. The students encountered a learning trajectory of procedures based on their physical constructions, process based on repeated procedures, and procept based on their reflections of the processes.

Weber found that experimental group of students performed much better in the post-test than the control group under traditional, text book driven lessons. There was evidence that the experimental students demonstrated a strong understanding since they considered trigonometric functions both as processes and the results of these processes, on which they could infer the properties of these functions. However, it is also clear that the experimental students were trained in this direction within the instructional design. For instance, one of the test items examined students' estimation of a sine value, and control group performed worse than the experimental group.

Considered that the experimental group performed such tasks in the lessons, and that such activities are not generally included in traditional methods, it is not surprising to find the experimental students outperforming the control students. Nevertheless, Weber's findings indicated that such an instructional design may help students develop a connected understanding.

Weber (2005) stated that the advantage of his unit circle trigonometry design was the process used by the students to create unit circle representations of the trigonometric functions, and that he did not think every unit circle method could lead to effective results. This idea confirms the findings of the research conducted by Kendal and Stacey (1997), who compared two methods of teaching trigonometry. They found that teaching the ratio method first promoted a better understanding of the trigonometric functions compared to the teaching starting with the unit circle method.

However, they assessed students' understanding with tasks of solving triangles. Therefore, it is expected that the ratio method would be more effective. It is important to note that the results of the investigations into students'. The notion of procept was defined by Gray and Tall (1994) as "the amalgam of the three components: a process, that produces a mathematical object, and a symbol which is used to represent either a process or an object" (as cited in Weber, 2005, p.92). Understanding depends largely on how the assessment is done.

Kendal and Stacey's study had a rather weak design in that they attempted to assess students understanding only through solving triangle tasks, which are actually not easily solvable using unit circle trigonometry. The research of Weber, and of Kendal and Stacey reveal that students' understanding is highly related to the aspects emphasized in the lesson design. Although the students were found to possess a deep understanding in certain conditions, it can be claimed that this understanding relied on the main focus of the teaching. The latter is consistent with the research of Delice and Roper (2006).

In a comparison study of English and Turkish students' performance on trigonometry, Delice and Roper found that Turkish students were good at algebraic aspects of trigonometry, such as simplifying trigonometric expressions, while English students were good at trigonometric applications in real life situations.

They mentioned that differences in mathematics curricula and in teaching methods in the two countries were the reason for this performance difference. Challenger (2009) discussed the same issue as "What you get is what you teach." It is difficult to find a firm framework for the understanding of trigonometry in the literature. Brown (2005) came up with a model of students' understanding of trigonometric functions, but focused on a very narrow side of trigonometry by including only the trigonometry of angles in degrees. Her model depicts that a student may consider the sine of an angle in three different ways: as a ratio, as a distance, and as a coordinate.

Brown showed that students who developed the most robust understanding were able to coordinate these different components. In this regard, a student with an integrated understanding is at the center of her triangular model of these three components. If a student cannot develop such an integrated view, he or she may have problems regarding certain trigonometric aspects. For instance, Brown found that students who had a distance-dominant understanding had difficulties with the sign of a trigonometric value, such as $\sin 210^0$.

Although Brown's model is useful for considering students' trigonometric understanding, it is far from being a model that can be used for a complete understanding of trigonometric functions. For instance, the model is based on the connections between triangle trigonometry and the unit circle, but does not tell us anything about the relationship between the unit circle and the graphs of trigonometric functions. The latter point seems crucial if students are to gain an integrated understanding of the different aspects of trigonometry. For this reason, a model which has a broader focus is proposed in the present study.

These two studies were the only studies found in the literature in which different teaching methods and students' understanding were explored in the same research study.

Trigonometry Instruction with Technology

Literature on technology-incorporated trigonometry teaching is also sparse and limited in scope. Although research shows positive effects of the use of technology on student learning and performance, possible effects of technology on student concept development remain rather unexplained. In an extensive search, no studies appeared to thoroughly examine the use of a technological tool in a way that directly supported student concept development of trigonometric functions which could be used to underpin the present study.

Blackett and Tall (1991) demonstrated that use of a computer software package addressing triangle trigonometry had a positive effect on student learning. The main advantage of this software compared to teaching without technology was that it enabled students to explore relationships between numeric and visual representations of trigonometric ratios in right triangles. Since then, there have been many changes in what technology can provide for students' learning. Kissane and Kemp (2009) exemplified the use of technology in the teaching of trigonometric concepts and procedures, ranging from calculators to computers. Calculators and computer programs based on computer algebra systems, and dynamic geometry software, all provide students with opportunities that are not available without technology, such as investigating multiple representations of functions in a dynamic way.

Regarding the use of computer algebra systems in the teaching of trigonometry, Stacey and Ball (2001) noted that calculators and computer packages, including computer algebra systems, could be used for solving triangles, trigonometric identities, and function properties. They also mentioned that using technology in solving triangles has the advantage of calculations with exact values. This is supported by Delice and Roper (2006), who found that English students using calculators to deal with real data, unlike Turkish students, were better in word problems. Johnson and Walker (2011) also mentioned the factor of exact values, and they found that students prefer to use a hand-held calculator, when the problems are difficult, to speed the process, and to verify their solutions. They also investigated how students make sense of mathematics by using technology, and concluded that technology helps students to make sense of challenging problems.

Johnson and Walker (2011) emphasized that technology can also be used for concept development with students, which seems to be another neglected area in the literature. In 2003, on the other hand, a study conducted by Choi-Koh investigated the effect of a graphing calculator on a single student's patterns of mathematical thinking. Choi-Koh (2003) reported that the student used the graphing calculator within three stages of mathematical thinking. First, he used it to graph the sine function, and to observe the changes in the graphs with varying coefficients in the intuitive stage. In the operative stage, the calculator then helped him explain trigonometric properties. Lastly in the applicative stage, the student used it to analyze and reflect on trigonometric relations.

Choi-Koh suggested that the movement between algebraic functions and their graphs was the main factor helping students proceed to the operative and applicative stages. Along with computer algebra systems, other researchers also investigated the effects of dynamic geometry software. Zengin, Furkan, and Kutluca (2012) compared GeoGebra-integrated constructivist teaching with solely constructivist teaching without GeoGebra, in a five-week trigonometry course. Although they did not provide enough information about their course design in their article, they reported that constructivist teaching with technology was more effective in terms of students' learning. They mentioned that GeoGebra is both a computer algebra system and dynamic geometry software, as it includes both symbolic and visualization features related to coordinates, equations and functions, along with geometric concepts and dynamic relations. Johari, Chan, Ramli, and Ahmat (2010) explored the effect of Geometer's Sketchpad on students' understanding of trigonometric functions in four categories: shape, maximum, minimum, and cycle.

As a weak point of their report, they were not explicit about what they meant by these four categories. They found that students using Geometer's Sketchpad demonstrated better understanding in the shape and cycle categories. This does not seem surprising if one considers that students probably do not need the support of a technological tool to learn the maximum and minimum values on the graph of the sine function. Johari et al.'s study also reveals that Geometer's sketchpad students did not show an understanding of drawing graphs. This finding could be directly related to design issues in their lessons, but it cannot be verified as they did not provide information about how they used technology in their design.

Using technology in lessons does not lead automatically to better results in terms of student learning and understanding. Of crucial importance are how technological tools are used in lessons, the kind of support students receive, and interactions between the tools and students. In relation to this, Thompson (2002) mentioned the use of technological tools as didactic objects that promote a reflective mathematics discourse for knowledge construction, but are not didactic when used alone. Moore (2009) used Geometer's Sketchpad as a didactic object enabling supportive mathematical discourse, as emphasized by Thompson. Moore used two applets for angle measurements and two covarying elements, namely arc length and vertical position, as a dynamic method for students, and found positive effects on students' understanding based on quantitative and covariational reasoning. He mentioned that such applets could lead to an understanding of the sine graph, which is not easy to draw because of concavity.

Moore's (2009) study indicates an effective use of technology in mathematics lessons. To increase effectiveness of technology for students' learning, technology must be integrated into classroom teaching in ways which will foster students' concept development and understanding, and address their learning difficulties. Ross, Bruce, and Sibbald (2011) investigated the placement of computer assisted learning in the teaching sequence as a factor which may also relate to its effectiveness.

They found that, in the case of transformations of the trigonometric functions, using a dynamic software package after whole class teaching of core concepts was more effective than beginning the learning unit with the software. Also, at the end of the learning unit there was almost no difference between the students' performance levels in the two groups. Hence, they suggested an integrated method of whole class teaching and technology use.

Related Subjects

Trigonometry is a complex subject and a deep understanding requires making connections and transitions among related concepts. In this regard, students' difficulties in trigonometric functions are attributed to the lack of understanding regarding the related areas, such as functions, angle measurements, and graphs. Gür (2009) mentioned that trigonometry and other related concepts are abstract and non-intuitive, and teaching based on lecturing does not help students overcome their learning difficulties. In this section, a literature review of the subjects related to trigonometry is presented. Note that this not only includes research findings specific to the education and understanding of trigonometry, but also theoretical considerations which constitute a basis for covering these subjects within the present study.

Angle Measurements

In general, trigonometry is first taught as ratios in right triangles. In this phase, it is limited to angles in degrees smaller than 90°. Subsequently, using a unit circle model, this scope is expanded to any angle and also angles in radians. A transition from radians to real numbers as the domain of trigonometric functions completes the picture. A rich trigonometric understanding requires being able to make transitions among these. Students' difficulties with angles can be considered as the most basic problem prohibiting them from developing a deep trigonometric understanding. Research shows that students have difficulties and an incomplete understanding regarding angles, both in degrees and in radians

Martinez-Sierra (2008) reported students' difficulties with the angle measure in degrees, particularly with negative angles and angles larger than 360°. Brown (2005) found that some students developed a fragile concept of rotation angles, hindering their performance in finding trigonometric ratios of angles due to problems with drawing rotation angles in the unit circle. Although students may have such difficulties with angles in degrees, research shows that students' trigonometry conception is mostly based on angles in degrees, and this may pose some other problems, such as difficulties with trigonometric values of real numbers or radians stated without multiples of π as trigonometric values of degrees. This is because students cannot develop a strong understanding of radians to be able to move to real numbers as the domain of trigonometric functions.

Along with students' difficulties with angles in degrees, Martinez-Sierra (2008) suggested that the use of radians is among the conceptual breaks in the construction of trigonometric functions. Akkoç and Akbaş Gül (2010) found that students could not define the radian concept as the ratio of two lengths nor relate angle measures to the lengths of subtending arcs. Orhun (2001) also reported students' difficulties to find the measure of arcs subtending given angles, and vice versa, noting that the calculation of angles from given arc lengths was more difficult for the students. Another student difficulty found in the literature relates to recognizing real numbers as radians.

Researchers found that students considered π radians as equal to 180, not as a number close to 3.14, and that such difficulties also exist in both pre-service and in-service teachers. They investigated concept maps of pre-service and in-service teachers and found that those with a stronger understanding of radians could use the unit circle and its connections to other trigonometric concepts. To address such problems with radians, some researchers suggested a way of teaching radians by demonstrating clear connections to the arc.

Akkoç and Akbaş Gül (2010), for instance, investigated a learning trajectory based on computer programs exploring the relationship between arc lengths and angles in different circles, and defining trigonometric functions, primarily in the domain of arc lengths. Their results show that such an approach can help students who have learning difficulties with the radian concept. Moore (2010) examined a teaching method that also emphasized the arc lengths in order to set the relationships between angles and arcs. He based his study on quantitative and covariational reasoning.

He suggested defining angles, both degrees and radians, as fractions of the circumference of a circle: 1 degree is the measure of an angle corresponding to the $(1/360)^{th}$ of any circle's circumference, while 1 radian corresponds to the $(1/2!)^{th}$ of any circle's circumference. These quantitative relations indicate how to interconnect degrees and radians, as well as relating angles to arc lengths and the radius of a circle. Moore (2010) found that such quantitative reasoning could constitute a fundamental understanding, and that investigating covariational changes of the arc lengths and vertical positions could be effective for students' understanding of trigonometric graphs based on radians.

Considering the importance of radians for students' trigonometric understanding, the present study put a great emphasis on this concept. To overcome students' difficulties, trigonometric functions were covered first through arc lengths. The transition from arcs to angles was expected to help students better understand the connection between arcs and angles, as well as the role of radians when transitioning to real numbers as the domain of trigonometric functions.

Function Concept

The most crucial aspect of trigonometry is perhaps related to the concept of a function. Research shows that the notion of a function is difficult in the context of trigonometry for students to conceptualize. Many studies on students' understanding of trigonometric functions considered trigonometry, based on the functions of angles only. However, analytic trigonometry based on trigonometric functions of real numbers is as important for students to understand as the trigonometry of angles. The former applies to the real periodic phenomena.

However, research on trigonometry appears to have ignored trigonometric functions of real numbers. Grabovskij and Kotel'nikov (1971) emphasized the significance of the trigonometric functions of numerical arguments, and advised the use of kinematic methods, such as angular velocity, as a way to avoid teaching trigonometric functions exclusively in terms of geometric ideas. Such a method can be effective in helping students to understand connections to certain real-life situations. However, it is probably not the most effective way to introduce them to analytic trigonometry due to difficulties related to physical, rather than mathematical, concepts. Teaching analytic trigonometry could be applied through radians as a way of connecting related algebraic and geometric underpinnings.

Research shows that this is not an easy process, and students have significant difficulties in conceptualizing trigonometric functions in the domain of real numbers. The study of Kang (2003) shows that 21% of the students claimed $f(x) = \sin x$ denotes the ratio of opposite to hypotenuse, and that 60% could not define $y = sin\,x$ as a function. Orhun (2001) also mentioned the difficulties of students within the area of trigonometry of numbers. For instance, the students considered since 30 as the sine of 30°.

Graphs and Cartesian Connection

The graphs of trigonometric functions are one of the most important aspects of trigonometric understanding. Research shows that students find the graphs of trigonometric functions difficult to grasp. Brown (2005) showed that students were unsuccessful at making even a very basic connection from the unit circle to the graph of the sine function, that is the transition between a position on the unit circle and the point that corresponds to it on the graph of y=sin x. Yet, such transitions are very important for a robust trigonometric understanding. Orhun (2001) advised teaching trigonometric functions through their graphs. Orhun's advice can be effective in addressing students' problems regarding the graphs of trigonometric functions and to transition from radians to real numbers.

The teaching of trigonometric functions in the present study started with an emphasis on the graph of the sine function. Graphs are important not only for trigonometric functions but also for other functions as they constitute one of the representations of a function. Connections between algebraic and graphical representations of functions are considered to be an important aspect of students' understanding, but such connections are not easy for most students to grasp. Thus, teaching graphs in a way that is connected to algebraic representations is advised by a number of researchers. Their study shows that students could not distinguish the graphs representing functions from those which do not. They also found that students were looking for continuity and a regularity in graphs in order to consider them as functions. Such issues may be due to a lack of a process conception of the function as well as being unable to utilize the formal definition.

An important framework for students' understanding regarding graphs based on connections between algebraic and graphical representations of functions is the Cartesian Connection. This includes two connections defined by Schoenfeld et al. (1993):

Connection A: The point (x_0, y_0) is on the graph of the function $y = f(x)$ if and only if the point satisfies the equation, that is $y_0 = f(x_0)$.

Connection B: In the Cartesian plane, specific algebraic expressions have graphical identities.

For example, $(y_2 - y_1)$ is a directed line segment with both direction and magnitude specified by mathematical convention? Knuth (2000) investigated students' abilities in terms of Connection A, only in the graph to the equation direction in linear functions, and found that students failed to recognize or create the connections based on the graphical representations.

For the case of trigonometry, these connections may be more challenging for students as a strong trigonometric understanding does not only require making connections between algebraic and graphical representations, but also making transitions between the unit circle and graphs, both of which are defined on the Cartesian plane. Connection A is important for students to recognize points like (x, cos x) on trigonometric graphs. Connection B, on the other hand, is important for realizing coordinates and line segments as trigonometric values defined with the help of the unit circle.

Periodicity

Periodicity is an important aspect of trigonometric functions that provides applications in explaining physical real-life phenomena, such as harmonic motion and waves. However, research on trigonometric functions has not examined periodicity although it is an important aspect of trigonometric understanding. From a mathematical point of view, the meaning of periodicity, and connections between trigonometric graphs and the unit circle with regards to periodicity are all important.

Since periodicity has not been studied in the context of trigonometry, two studies from a general mathematical stance seem important for the present study. In one of these studies, Van Dormolen and Zaslavsky

(2003) discussed the issues revealed by some definitions of periodicity. Two of them are as follows:

Definition A: A function f is called periodic if there exists a non-zero number p, such that for every x that belongs to the domain of f, the following conditions are fulfilled:

(a) x ± p belongs to the domain of f,

(b) f(x ± p) = f(x).

Definition B: A function is called periodic if there exists a non-trivial translation of the graph of f along the horizontal axis, such that the image coincides with the original.

Van Dormolen and Zaslavsky mentioned that these two definitions are equivalent, but definition A is analytic whereas definition G is global. They also noted that A is taken from a point-wise perspective while B relates to the function as a whole.

UNITS OF AN ANGLE

Concept of an angle

Two rays with a common starting point from an angle. One of the rays of the angle is the initial side and the terminal side. The angle is identified by showing the direction of the rotation from the initial side to the terminal side.

An angle is said to be positive/negative if the rotation is anti-clockwise/clockwise. Angles are usually denoted by Greek letters such as α (alpha), β (beta), γ (gamma), θ (theta) etc.

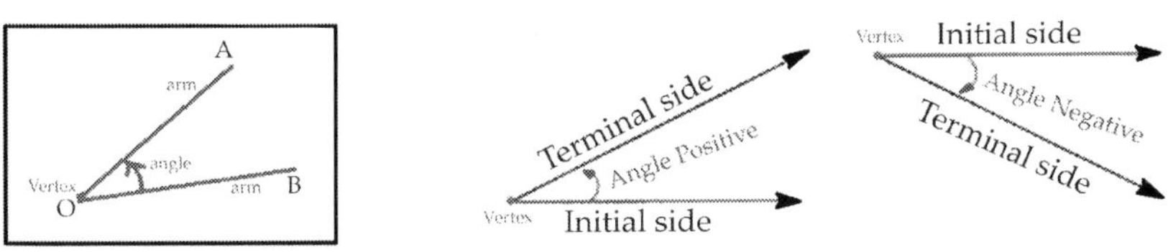

Figure 1.1

There are two commonly used measurements for angles; degrees and radians. These are explained as below.

Sexagesimal System (or English System)

Sexagesimal is a numeral system with 60 at its base. It originated with the ancient Sumerians in the 3rd millennium BC, was passed down to the ancient Babylonians, and is still used in a modified form for measuring time, angles and geographic coordinates. If the initial ray rotates in an anti-clock wise direction in such a way that it coincides with itself, the angle then formed is said to be of 360°.

In the sexagesimal system, an angle is measured in degrees, minutes and seconds. A complete rotation describes 360^0. In this system, a right angle is divided into 90 equal parts and each part is denoted as a degree 1°. A degree is then divided into 60 equal parts and such part is denoted as a minute $1'$, and a minute is then divided into 60 equal parts, each of which is called a second $1''$.

In short, we can write:

1 rotation (anti –clock wise) $= 360^\circ$

One degree (1^0) $= 60'$

One Minute ($1'$) $= 60''$.

Circular System

The circular system is most useful for the study of higher mathematics. Particularly in calculus, angles are measured in radians. In this system, a radian is considered as the unit for the measurement of angles.

Radian: Radian is the measure of the angle subtended at the center of the circle by an arc, whose length is equal to the radius of the circle. In any circle, the angle subtended at its center by an arc of a circle whose length is equal to the radius of the circle is called a radian. Let $OX = r$ be the radius of a circle having a center O.

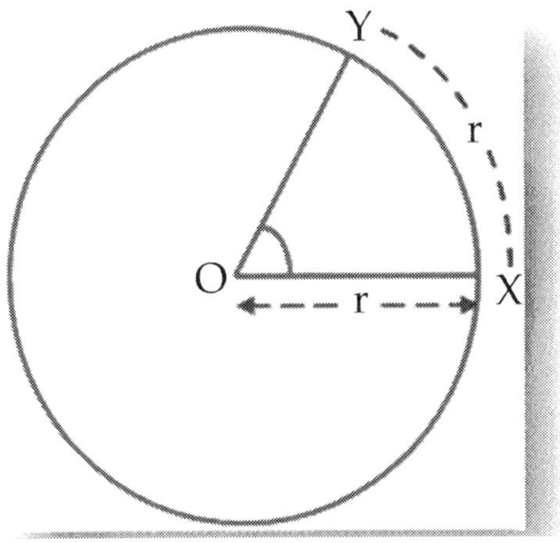

Figure 1.2

This brings us to the next system in trigonometry; the centesimal system (or French system). In this system, an angle is measured in grades, minutes and seconds. A right angle is divided into 100 equal parts, known as grades. A grade is divided into 100 equal parts and each part is called a centesimal minute. A minute is further divided into 100 equal parts, known as centesimal seconds.

We can see that minute and seconds in sexagesimal and centesimal systems are different.

Conversion from degrees minutes seconds to a decimal form and vice versa.

If we want to convert degrees minutes seconds to decimal form then we will use following method.

Example 1:

$16^\circ 30' = 16.5^\circ$ as we can write $30' = \left(\frac{1}{2}\right)^\circ = 0.5^\circ.$

Similarly, we can write

$45.25^\circ = 45^\circ 15'$, as we can write $\left(0.25 = \frac{25}{100} = \frac{1}{4} = \frac{60}{4} = 15\right).$

Example 2: Convert $18^\circ 6' 21''$ to decimal form.

Solution: Write $18^{\circ}6^{'}21^{''}=\left[18+6\left(\frac{1}{60}\right)+21\left(\frac{1}{60\times60}\right)\right]$

$=\left(18+0.1+0.005833\right)^{\circ}$

$=18.105833^{\circ}$.

Example 3: Convert 21.256° to the D M S form

Solution: $0.256^{\circ}=\left(0.256\right)\times\left(1^{\circ}\right)$

$=\left(0.256\right)\times\left(60^{'}\right)$

$=15.36^{'}$.

And now consider $0.36^{'}=\left(0.36\right)\times\left(1'\right)$

$=\left(0.36\right)\times\left(60^{''}\right)=21.6^{''}$.

Therefore,

$21.256^{\circ}=21^{\circ}+0.256^{\circ}$

$=21^{\circ}+15.36^{'}$

$=21^{\circ}+15^{'}+0.36^{''}$

$=21^{\circ}+15^{'}+21.6^{''}$

$=21^{\circ}15^{'}22^{''}$

RELATIONSHIP BETWEEN THE LENGTH OF AN ARC AND THE CIRCULAR MEASURE OF AN ANGLE

Theorem: Prove that $\theta=\frac{l}{r}$, where r is the radius of a circle, l is the length of the arc and θ is circular measure of the central angle.

Proof:

Let there be a circle with center O and radius r. Suppose that length of arc $\widehat{AB} = l$ and the central angle $m < AOB = \theta$ radian. Take an arc $\widehat{AC}$ of length r.

Now by definition $m < AOC = 1$ radian.

We know from elementary geometry that measures of central angles of the arcs of a circle are proportional to the lengths of their arcs.

$$\Rightarrow \frac{m < AOB}{m < AOC} = \frac{m\ \widehat{AB}}{m\ \widehat{AC}}$$

$$\Rightarrow \frac{\theta\, radian}{1\, radian} = \frac{l}{r}$$

$$\Rightarrow \theta = \frac{l}{r}.$$

Thus, the central angle θ (in radians) subtended by a circular arc of length l is given by $\theta = \frac{l}{r}$, where r is the radius of the circle.

Note: Using this formula remember these points.

- Radius (r) and arc length (l) should be of the same unit i.e. if the arc length is in meters than the radius also is in meters.
- Always use θ in radian form. If it is given in degree then convert it to radians first.

Angular Velocity

The relationship between the linear velocity v and the angular velocity ***o*** (the Greek letter ***omega)*** for an object with radius ***r*** is $v = r\omega$, where ω is measured in radians per unit of time and v is distance per unit of time.

(NOTE: **r** and ω use the same unit of time and ***r*** and ***v*** use the same linear unit.)

CONVERSION OF DEGREES INTO RADIANS AND VICE VERSA

We will use formula $1^0 = \frac{\pi}{180}\ radian$ to convert degrees into radians. Similarly, for the conversion of radians to degrees we will apply the formula $1\ radian = \frac{180}{\pi}$.

Here are some examples of how we can apply both formulas to convert degrees to radians and vice versa.

Example 4: Convert the following sexagesimal measure of angles into radians.

- 150°
- $75^{\circ}\,6'30''$
- $10^{\circ}\,15'$.

Solution:

- We know that $1^0 = \frac{\pi}{180}\ radian$,

hence, $150^{\circ} = 150 \times \frac{\pi}{180}$

$= \frac{5\pi}{6}\ radian$

- Firstly, we convert degrees minutes and seconds to decimal form using the method mention above:

$= 75^{\circ} + \frac{6}{60} + \frac{30}{3600}$

$= 75.1083^{\circ}$.

So now apply formula $1^0 = \frac{\pi}{180}\ radian$:

$75.1083^{\circ} = 75.1083 \times \frac{\pi}{180} = 1.3109.$

- Firstly, we convert degrees minutes to decimal form:

$10^{\circ}15' = 10 + \frac{15}{60} = 10.25^{\circ}$

Apply the formula:

$$10.25^\circ = 10.25 \times \frac{\pi}{180} = 0.1789\, radians$$

Example 5: Convert the following angles in radians into measures of the sexagesimal system.

i. $\frac{\pi}{8} radian$

ii. $\frac{13}{16}\pi\, radian$

iii. $\frac{2}{3}\pi\, radian.$

Solution:

i. We know that $1\, radian = \frac{180}{\pi}$,

ii. Now $\frac{\pi}{8} radian = \frac{\pi}{8} \times \frac{180}{\pi} = 22.5^\circ$.

iii. $\frac{13}{16}\pi$

$$\frac{13}{16}\pi\, radian = \frac{13}{16}\pi \times \frac{180}{\pi} = 146.25^\circ.$$

iv. $\frac{2}{3}\pi = \frac{2}{3}\pi \times \frac{180}{\pi} = 120^\circ.$

Applications of formula $l = r\theta$

This formula is used in different type of problems. We present various practical problems to show how important this formula is.

Problem 1. What is the length of the arc intercepted on a circle of radius 14 cm by the arms of a central angle of 45° ?

Solution: We have $r = 14$ cm, $\theta = 45^\circ$,

Firstly, we will convert angle θ into radians by the method already discussed :

$\theta = \frac{\pi}{4}, r = 14\,\text{cm}$. Now we have to find length of arc i.e. l:

Apply formula $l = r\theta$

$l = 14 \times \frac{\pi}{4} = 10.9956$ cm.

Problem 2. A railway train is running on a circular track of radius 500 m at the rate of 30 km per hour. Through what angle will it turn in 10 secs?

Solution. The given data is $r = 500$ m

Speed = 30 km per hr.

Firstly, we will convert km/h to m/s, as the radius is given in meters.

$$\text{Speed} = \frac{30 \times 1000}{3600} = \frac{25}{3}\ m/s.$$

Distance in 1 sec is = $\frac{25}{3} m$

Distance in 10 secs $= l = \frac{25}{3} \times 10 = \frac{250}{3} m$

Apply formula $\frac{l}{r} = \theta$

$$\theta = \frac{\frac{250}{3}}{500} = \frac{1}{6}\ radian.$$

Problem 3. A horse is tethered to a peg by a rope of 27 m length and it can move in a circle with the peg as center. If the horse moves along the circumference of the circle, keeping the rope tight, how far will it have gone when the rope has turned through an angle of 70°?

Solution: In this problem, the given data is $r = 27m$ and $\theta = 70^\circ$.

Firstly, we will convert θ into radian form:

$$\theta = 70^\circ = 70 \times \frac{\pi}{180}\ \text{rad} = \frac{7\pi}{18}\ \text{rad}$$

r = 27m; Hence, we have to find the value of the arc length l.

Apply the formula:

$$l = r\theta$$

$$l = \frac{7\pi}{18} \times 27 = 10.5\pi = \mathbf{33m}.$$

Problem 4. How far apart are two cities on the equator whose longitudes are 10° E and

50° W? (Radius of the Earth is 6400 km).

Solution: $\theta = 10° + 50° = 60°$

$$\theta = 60 \times \frac{\pi}{180} \text{rad} = \frac{\pi}{3} \text{ rad}$$

r = radius of Earth = 6400 km , as we know that

$$\mathbf{l = r\theta}$$

$$= 6400 \times \frac{\pi}{3}$$

$l = 6704.76$ km .

Problem 5. A man land on the moon and observes that the Earth's distance subtends an angle of $1^\circ 54'$ at his place of landing. If the Earth's radius is 6400 km , Find the distance between the Earth and the moon.

Solution: Let A and B be the center of Earth and moon respectively.

$r = |AB| = ?, \quad l = |CD| = 2 \times 6400 = 12800$ km

$$\theta = 1°54' = 1.9 \times \frac{\pi}{180} = 0.03316 \text{ rad}$$

$$r = \frac{l}{\theta} = \frac{12800}{0.03316} = 386240 \text{ km}.$$

Problem 6. The sun is about 1.496×10^8 km away from the Earth. If the angle subtended by the sun on the surface of the Earth is 9.3×10^{-3} radians approximately. What is the distance of the sun from the Earth?

Solution: Let A and B be the center of Earth and moon respectively.

$r = |AB| = 1.496 \times 10^8$ km, $\quad \theta = 9.3 \times 10^{-3}$ rad , $l =$ diameter of sun = ?

$l = r\theta$

$l = 1.496 \times 10^8 \times 9.3 \times 10^{-3}$

$= 1391280$ km

$l = 1.39 \times 10^6$ km.

Problem 7. A circle wire of radius 6 cm is cut, straightened and then bent to lie along the circumference of a hoop of radius 24 cm. Find the measure of the angle that subtends at the center of the hoop.

Solution:

$$l = \text{circumference of circular wire of radius } 6\,\text{cm} = 2\pi(6) = 12\pi\,\text{cm}$$

$$r = 24\,cm \;;\; \theta = ?$$

$$\theta = \frac{l}{r}.$$

$$= \frac{12\pi}{24}$$

$$= \frac{\pi}{2}\,rad$$

Problem 8. A pendulum of 5 m long swings through an angle of 4.5°. Through what distance does the bob move?

Solution: $\theta = 4.5° = 4.5 \times \frac{\pi}{180} rad = 0.07854\,rad$

$$r = 5m \;;\; l = ?$$

$$l = r\theta$$

$$l = 0.07854 \times 5 = 0.39\,m.$$

Problem 9. A flywheel rotates at 300 rev/min. If the radius is 6 cm, through what total distance does a point on the rim travel in 30 secs?

Solution: $\omega = 300\,rpm \;\; ; t = 30\,sec = \frac{30}{60} = \frac{1}{2} = 0.6\,min$

$$\theta = 2\pi\,\omega t = 2\pi \times 300 \times 0.6 = 300\pi\ rad$$

$$r = 6\,cm$$

$$l = r\theta = 6 \times 300\pi = 1800\pi = 5654\,cm.$$

AREA OF A SECTOR OF A CIRCULAR REGION

Proof: Consider a circle with radius r, angle θ, and denote the area of a sector with A.

Since we know from elementary geometry that

$$\frac{\text{Area of Sector}}{\text{Area of Circle}} = \frac{\text{Angle of Sector}}{\text{Angle of Circle}}$$

Area of Sector = A

Area of Circle = πr^2

Angle of Sector = θ

Angle of circle = 2π

We can apply,

$$\frac{A}{\pi r^2} = \frac{\theta}{2\pi}.$$

After simplying, we will obtain

$$A = \frac{1}{2} r^2 \theta$$

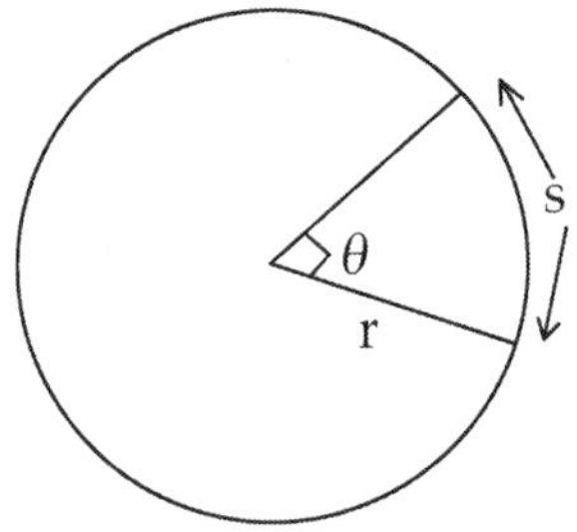

ANGLE IN STANDARD POSITION

An angle is said to be in a standard position if its vertes lies at the origin of a rectangular coordinate system and its initial side along the positive x-axis.

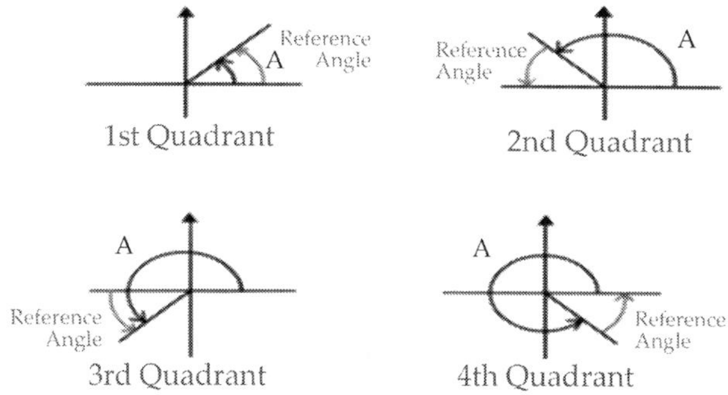

Figure 1.4

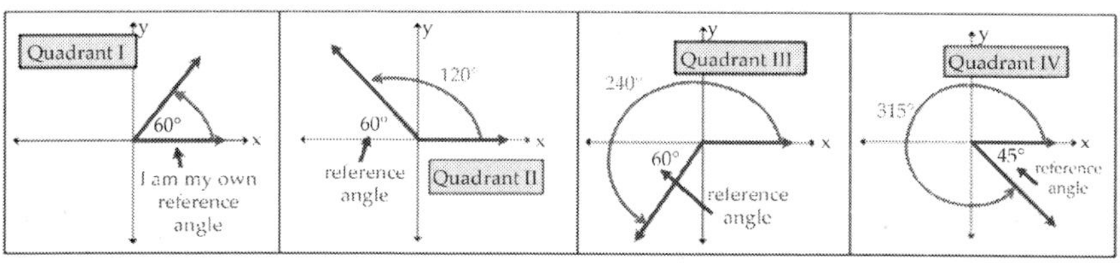

Figure 1.5

Coterminal angle

There can be many angles with the same initial and terminal sides. These are called coterminal angles.

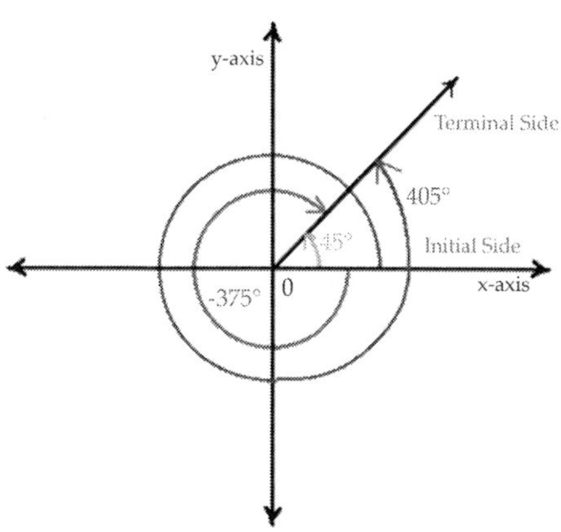

Figure 1.6

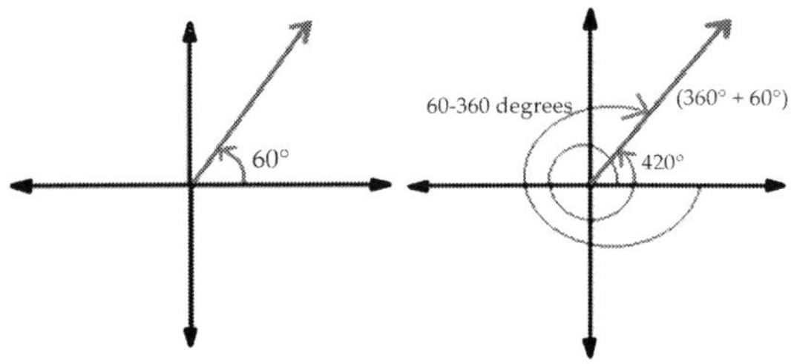

Figure 1.7

In general, if angle θ is in degress, then $\theta + 360k$, where $k \in Z$, is an angle coterminal with θ. If angle θ is in radians, then $\theta + 2k\pi$ where $k \in Z$, is an angle coterminal with θ.

So we can say that in general, the angle is $\theta + 2k\pi$, $k \in \mathbf{Z}$.,

Coordinate plane

A rectangular coordinate system in a plane consists of a horizontal and vertical number scale (called axes), whose point of intersection (origin) is the origin on each scale. It is customary to choose the positive direction on each axis. that is, positive to the right on the horizontal axis or x-axis and positive upward on the vertical or y-axis. For convenience, we shall assume the same unit of measure on each axis.

By means of such a system, the position of any point P in the plane is given by its (directed) distances, called coordinates, from the axes. A point P with abscissa x and ordinate y is denoted by P(x, y). The axes divide the plane into four parts, called quadrants, which are numbered in a counterclockwise direction as I, 11, 111, and IV.

Concept of quadrants

Each of four quarters of a circle are denoted a quadrants. When we divide a circle into four equal parts, we get four quadrants, as shown in the figure below:

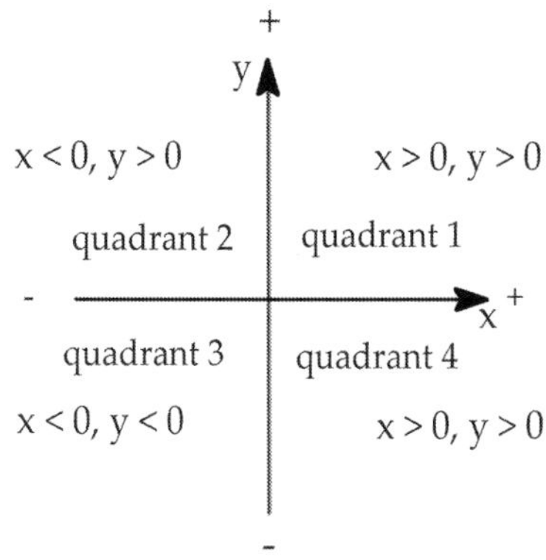

Figure 1.8

All Students Take Care

In the 1st quadrant, all trignometric ratios will be positive.

In the 2nd quadrant, $Sin\,\theta$, and $Cosec\,\theta$ will be positive, and all other trignometric ratios will be negative.

In the 3rd quadtrant, all trigonometric ratios will be negative, apart from $\tan\theta$ and $\cot\theta$.

In the 4th quadrant, $\cos\theta$, and $\sec\theta$ will be positive, and all other trigonometric ratios will be negative.

The angle division of all quadrants are

In the 1st quadrant, angles will be in the range of $0 \leq \theta \leq 90$

In the 2nd quadrant, angles will be in the range of $90^0 \leq \theta \leq 180^\circ$

In the 3rd quadrant, angles will be in the range of $180^0 \leq \theta \leq 270^\circ$

In the 4th quadrant, angles will be in the range of $270^0 \leq \theta \leq 360^\circ$

TRIGNOMETRIC FUNCTIONS

Consider a right angle triangle, with right angle at point B.

Suppose we take angle A then:

The side AC opposite to the 90° is called the hypotenuse.

The side BC opposite to angle A is called the perpendicular or opposite.

And sthe ide AB is called the base or adjacent.

Hence, we can form six ratios:

$Sin\ A = \frac{opp}{hyp} = \frac{a}{b}$; $Cosec\ A = \frac{hyp}{opp} = \frac{b}{a}$

$Cos\ A = \frac{adj}{hyp} = \frac{c}{b}$ $Sec\ A = \frac{hyp}{adj} = \frac{b}{c}$

$Tan\ A = \frac{opp}{adj} = \frac{a}{c}$ Cot $A = \frac{adj}{opp} = \frac{c}{a}$

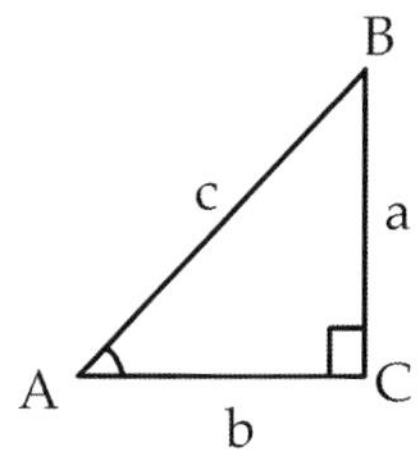

Figure 1.9

Undefined trigonometric functions

It has been noted that $\cot 0^\circ$ and $\csc 0^\circ$ are not defined since division by zero is not allowed in mathematics, but the values of these fuctions for angles near 0° are of interest. If we take θ to be a very small positive near to zero then we see that $cot\theta \to +\infty$ and $\csc\theta \to +\infty$.

Now again, if we take θ to be a negative very near to zero, we will observe that $cot\theta \to -\infty$ and $\csc\theta \to -\infty$.

Similarly, different trigonometric functions are undefined at different angles, as seen in the table below.

Angle θ	Function Value
$\theta \to 0^\circ$	$\cot\theta \to \pm\infty$, $\csc\theta \to \pm\infty$

$\theta \to 90^\circ$	$\tan\theta \to \pm\infty$, $\sec\theta \to \pm\infty$
$\theta \to 180^\circ$	$\cot\theta \to \pm\infty$, $\csc\theta \to \pm\infty$
$\theta \to 270^\circ$	$\tan\theta \to \pm\infty$, $\sec\theta \to \pm\infty$

FUNDAMENTAL IDENTITIES

For any real number θ, we shall derive the following three fundamental identities:

$\sin^2\theta + \cos^2\theta = 1$

$1 + \tan^2\theta = \sec^2\theta$

$1 + \cot^2\theta = cosec^2\theta$

- Consider a right triangle, as shown in the figure.

Now by Pythagoras theorem, we have,

dividing $a^2 + b^2 = c^2$ both sides by c^2, we get

$$\frac{a^2}{c^2} + \frac{b^2}{c^2} = \frac{c^2}{c^2}.$$

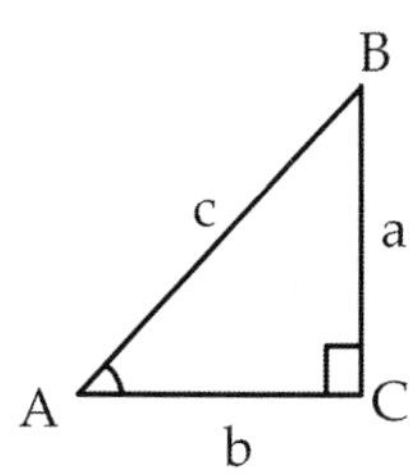

Figure 1.10

$$\Rightarrow \quad \left(\frac{a}{c}\right)^2 + \left(\frac{b}{c}\right)^2 = 1$$

$$\Rightarrow \quad (\sin\theta)^2 + (\cos\theta)^2 = 1$$

hence, prove that $\sin^2\theta + \cos^2\theta = 1$.

Again as $a^2+b^2=c^2$,

dividing both sides by b^2, we get:

$$\frac{a^2}{b^2}+\frac{b^2}{b^2}=\frac{c^2}{b^2}.$$

$$\Rightarrow \quad \left(\frac{a}{b}\right)^2+1=\left(\frac{c}{b}\right)^2$$

$$\Rightarrow \quad (\tan\theta)^2+1=(\sec\theta)^2$$

$$\Rightarrow \quad 1+\tan^2\theta=\sec^2\theta.$$

- Again as $a^2+b^2=c^2$,

dividing both sides by a^2, we get

$$\frac{a^2}{a^2}+\frac{b^2}{a^2}=\frac{c^2}{a^2}$$

$$\Rightarrow \quad 1+\left(\frac{b}{a}\right)^2=\left(\frac{c}{a}\right)^2$$

$$\Rightarrow \quad 1+(\cot\theta)^2=(\operatorname{cosec}\theta)^2$$

$$\Rightarrow \quad 1+\cot^2\theta=\operatorname{cosec}^2\theta.$$

Signs of the trigonometric functions

If θ is not a quadrantal angle, then it will lie in a particular quadrant. As $r=\sqrt{x^2+y^2}$ is always positive, it follows that the signs of the trigonometric functions can be found if the quadrant of θ is known. For example:

- If θ lies in quadrant I, then a point $P(x, y)$ on its terminal side has both x and y co-ordinates positive.
  $\Rightarrow$ All trigonometric functions are positive in the 1st quadrant.

- If θ lies in quadrant II, then a point $P(x, y)$ on its terminal side has negative x-coordinates and positive y-coordinates.

$$\therefore \sin\theta = \frac{y}{r} > 0(positive)\ ,Cos\,\theta = \frac{x}{r} < 0\ (negative)$$

$$tan\theta = \frac{y}{x} < 0(\,negative)$$

- If θ lies in quadrant III, then a point $P(x, y)$ on its terminal side has negative x-coordinates and negative y-coordinates.

$$\therefore \sin\theta = \frac{y}{r} < 0(negative)\ ,Cos\,\theta = \frac{x}{r} < 0\ (negative)$$

$$tan\theta = \frac{y}{x} > 0(\,positive)$$

- If θ lies in quadrant IV, then a point $P(x, y)$ on its terminal side has positive x-coordinates and negative y-coordinates.

$$\therefore \sin\theta = \frac{y}{r}\left\langle 0(negative)\ ,Cos\,\theta = \frac{x}{r}\right\rangle 0\ (positive)$$

$$tan\theta = \frac{y}{x} < 0(\,negative)$$

These results are summarized in the following figure.

II Quadrant	1st Quadrant
$\begin{cases}\sin\theta > 0\\ \csc\theta > 0\end{cases}$	All + ve
III Quadrant	IV Quadrant
$\begin{cases}\tan\theta > 0\\ \cot\theta > 0\end{cases}$	$\begin{cases}\cos\theta > 0\\ \sec\theta > 0\end{cases}$

It is clear from the above figure that

$$\sin(-\theta) = -Sin\,\theta; \qquad csc(-\theta) = -Csc\,\theta$$

$$\cos(-\theta) = cos\,\theta; \qquad sec(-\theta) = sec\,\theta$$

$\tan(-\theta) = -\tan\theta$; $\quad cot(-\theta) = -cot\,\theta$

Problem 1: If $\tan\theta = \frac{8}{15}$ and the terminal arm of the angle is in quadrant III, find the values of the other trigonometric functions of θ.

Solution: Given that $\tan\theta = \frac{8}{15}$, we have $\cot\theta = \frac{15}{8}$.

Apply identity $1 + \tan^2\theta = \sec^2\theta$

$$\sec^2\theta = 1 + \left(\frac{8}{15}\right)^2 = 1 + \frac{64}{225} = \frac{289}{225}$$

$$\therefore \sec\theta = \pm\sqrt{\frac{289}{225}}$$

$$\sec\theta = \pm\frac{17}{15}.$$

Now, the terminal arm of the angle is in the III quadrant, where $\sec\theta$ is negative, hence,

$$\sec\theta = -\frac{17}{15}.$$

Now $\cos\theta = \frac{1}{sec\theta} = -\frac{15}{17}$

$$\sin\theta = tan\theta\ .\cos\theta = \frac{8}{15} \times \left(-\frac{15}{17}\right)$$

$$\sin\theta = -\frac{8}{17}$$

and $\csc\theta = \frac{1}{sin\theta} = -\frac{17}{8}$

Problem 2: Find the value of five trigonometric functions of θ, if $\cos\theta = \frac{12}{13}$ and the terminal side of the angle is not in quadrant I.

Solution : Since the terminal side of the angle is not in quadrant I, but $\cos\theta$ is positive,

the terminal side of the angle is in the IV quadrant.

Now $\sec\theta = \dfrac{1}{\cos\theta} = \dfrac{13}{12}$

$$\sin^2\theta = 1 - \cos^2\theta = 1 - \left(\frac{12}{13}\right)^2$$

$$= 1 - \frac{144}{169} = \frac{25}{169}$$

$$\therefore \quad \sin\theta = \pm\frac{5}{13}.$$

As the terminal side of the angle is in the IV quadrant, where $\sin\theta$ is negative.

$$\therefore \quad \sin\theta = -\frac{5}{13}$$

$$cosec\,\theta = \frac{1}{\sin\theta} = -\frac{13}{5}$$

$$\tan\theta = \frac{\sin\theta}{\cos\theta} = -\frac{5}{12}.$$

Problem 3. If If $\cos\theta = -\frac{\sqrt{3}}{2}$ and the terminal side of the angle lies in the third quadrant, find the remaining trigonometric ratios of θ.

Solution. $AB = \sqrt{3};\ AC = 2$

By Pythagorean Theorem:

$$(AC)^2 = (AB)^2 + (CA)^2$$

$$(2)^2 = \left(\sqrt{3}\right)^2 + (CA)^2 \Rightarrow (CA)^2 = 3 - 2$$

$$CA = \pm\sqrt{1} \Rightarrow \mathbf{CA} = 1$$

$\sin\theta = -\frac{1}{2}$; $\text{cosec}\theta = -2$

$\cos\theta = -\frac{\sqrt{3}}{2}$; $\sec\theta = -\frac{2}{\sqrt{3}}$

$\tan\theta = \frac{1}{\sqrt{3}}$; $\cot\theta = \sqrt{3}$

Problem 4. If $\sin\theta = \frac{3}{8}$ and the terminal side of the angle lies in the second quadrant, find the remaining trigonometric ratios of θ.

Solution: $BC = 3\ ; AC = 8$

By Pythagorean Theorem:

$$(AC)^2 = (AB)^2 + (BC)^2$$

$$(8)^2 = (AB)^2 + (3)^2$$

$$(AB)^2 = 64 - 9$$

$$AB = \sqrt{55}$$

$\sin\theta = \frac{3}{8}$; $\text{cosec}\theta = \frac{8}{3}$

$\cos\theta = -\frac{\sqrt{55}}{8}$; $\sec\theta = -\frac{8}{\sqrt{55}}$

$\tan\theta = -\frac{3}{\sqrt{55}}$; $\cot\theta = -\frac{\sqrt{55}}{3}$.

Problem 5. If If $\sin\theta = \frac{2}{3}$ and the terminal side of the angle lies in the second quadrant, find the remaining trigonometric ratios of θ.

Solution: $BC = 3\ ; AC = 3$

By Pythagorean Theorem:

$$(AC)^2 = (AB)^2 + (BC)^2$$

$$(3)^2 = (AB)^2 + (2)^2$$

$(AB)^2 = 9 - 4$

$AB = \sqrt{5}$

$\sin\theta = \frac{2}{3}$; $\text{cosec}\theta = \frac{3}{2}$

$\cos\theta = -\frac{\sqrt{5}}{3}$; $\sec\theta = -\frac{3}{\sqrt{5}}$

$\tan\theta = -\frac{2}{\sqrt{5}}$; $\cot\theta = -\frac{\sqrt{5}}{2}$

Problem 6. If If $\tan\theta = \frac{3}{4}$ and the terminal side of the angle lies in the second quadrant, find the remaining trigonometric ratios of θ.

Solution: $BC = 3$; $AB = 4$

By Pythagorean Theorem:

$(AC)^2 = (AB)^2 + (BC)^2$

$(AC)^2 = (4)^2 + (3)^2$

$(AC)^2 = 16 + 9$

$\mathbf{AC} = 5$

$\sin\theta = -\frac{3}{5}$; $\text{cosec}\theta = -\frac{5}{3}$

$\cos\theta = -\frac{4}{5}$; $\sec\theta = -\frac{5}{4}$

$\tan\theta = \frac{3}{4}$; $\cot\theta = \frac{4}{3}$.

Problem 7. If $\tan\theta = -\frac{1}{3}$ and the terminal side of the angle lies in the second quadrant, find the remaining trigonometric ratios of θ.

Solution: $BC = 1$; $AB = 3$

By Pythagorean Theorem:

$$(AC)^2 = (AB)^2 + (BC)^2$$

$$(AC)^2 = (3)^2 + (1)^2$$

$$(AC)^2 = 9 + 1$$

$$AC = \sqrt{10}$$

$\sin\theta = \frac{1}{\sqrt{10}}$; $\operatorname{cosec}\theta = \sqrt{10}$

$\cos\theta = -\frac{3}{\sqrt{10}}$; $\sec\theta = -\frac{\sqrt{10}}{3}$

$\tan\theta = -\frac{1}{3}$; $\cot\theta = -3$

Problem 8. If $\cot\theta = \frac{4}{3}$ and the terminal side of the angle does not lie in the first quadrant, find the remaining trigonometric ratios of θ.

Solution: $BC = 3$; $AB = 4$

By Pythagorean Theorem:

$$(AC)^2 = (AB)^2 + (BC)^2$$

$$(AC)^2 = (4)^2 + (3)^2$$

$$(AC)^2 = 16 + 9$$

$$AC = 5$$

$\sin\theta = -\frac{3}{5}$; $\operatorname{cosec}\theta = -\frac{5}{3}$

$\cos\theta = -\frac{4}{5}$; $\sec\theta = -\frac{5}{4}$

$\tan\theta = \frac{3}{4}$; $\cot\theta = \frac{4}{3}$

Problem 9. If $\cot\theta = \frac{2}{3}$ and the terminal side of the angle does not lie in the first quadrant, find the remaining trigonometric ratios of θ.

Solution: $BC = 3\ ; AB = 2$

By Pythagorean Theorem:

$$(AC)^2 = (AB)^2 + (BC)^2$$

$$(AC)^2 = (2)^2 + (3)^2$$

$$(AC)^2 = 4 + 9$$

$$AC = \sqrt{13}$$

$$\sin\theta = -\frac{3}{\sqrt{13}} \quad ; \quad \text{cosec}\theta = -\frac{\sqrt{13}}{3}$$

$$\cos\theta = -\frac{4}{\sqrt{13}} \quad ; \quad \sec\theta = -\frac{\sqrt{13}}{4}$$

$$\tan\theta = \frac{3}{2} \quad ; \quad \cot\theta = \frac{2}{3}$$

Problem 10. If $\sin\theta = \frac{4}{5}$ and $\frac{\pi}{4} < \theta < \pi$, find the trigonometric ratio of $\boldsymbol{\theta}$.

Solution: As θ lies in 2nd quadrant, so

$$BC = 4\ ; AC = 5$$

By Pythagorean Theorem:

$$(AC)^2 = (AB)^2 + (BC)^2$$

$$(5)^2 = (AB)^2 + (4)^2$$

$$(AB)^2 = 25 - 16$$

$$AB = 3$$

$\sin\theta = \frac{4}{5}$; $\operatorname{cosec}\theta = \frac{5}{4}$

$\cos\theta = -\frac{3}{5}$; $\sec\theta = -\frac{5}{3}$

$\tan\theta = -\frac{4}{3}$; $\cot\theta = -\frac{3}{4}$

Problem 11. If $\sin\theta = \frac{7}{25}$ find $\cos\theta$ if angle θ is an acute angle.

Solution: $BC = 7;\ AC = 25$

By Pythagorean Theorem:

$$(AC)^2 = (AB)^2 + (BC)^2$$

$$(25)^2 = (AB)^2 + (7)^2$$

$$(AB)^2 = 625 - 49$$

$$AB = \sqrt{576}$$

$$AB = 24$$

$$\cos\theta = \frac{24}{25}$$

Problem 12. If $\sin\theta = \frac{5}{6}$, find $\cos\theta$ if angle θ is an obtuse angle.

Solution: Since angle θ lies in 2nd quadrant, so

$BC = 5\ ;AC = 6$

By Pythagorean Theorem:

$$(AC)^2 = (AB)^2 + (BC)^2$$

$$(6)^2 = (AB)^2 + (5)^2$$

$$(AB)^2 = 36 - 25$$

$$AB = \sqrt{11}$$

$$\cos\theta = -\frac{6}{\sqrt{11}}$$

SUMMARY

A triangle is the most vital form in our learning of mathematics. The word trigonometry means the measurement of triangles. The study of the sun, Earth, and the other planets has advanced by an understanding of the ratios of the sides of similar triangles. Eratosthenes (276–194 B.C.) used right angle triangles to estimate the circumference of the Earth to about 25,000 miles. If we compare this to the best modern estimate, 24,902 miles, we see that even though his methods involved some inaccuracies, his final results were outstanding. Although in the history of mathematics, the applications of trigonometry are based on right angle triangles, the scope of trigonometry is much greater than that. Today, trigonometry is vital to fields ranging from computer art to satellite communications.

Many machines drive by revolving a wheel. As a wheel turns, each spoke from the center of the wheel moves through a central angle and any point on the rim of the wheel moves through an arc whose degree measure is equal to that of the central angle. The angle and the arc can have any real number as their measure. In this chapter, we cover numerous topics, starting from the definition of trigonometry, the history of trigonometry and concepts of an angle in degrees and radians. Following these, concepts of trigonometric ratios and trigonometric functions and then their application using various problems are presented.

Review of common terms and basic formulas

Adjacent side: In a right triangle, the side between a given angle and the right angle.

Angle: Two rays with a common starting point from an angle. One of the rays of angle is called the initial side and the other the terminal side. The angle is identified by showing the direction of rotation from the initial side to the terminal side.

Radian: Radian is the measure of the angle subtended at the center of the circle by an arc, whose length is equal to the radius of the circle.

Angular speed: The angle through which a rotating object travels in a unit of time.

Angular velocity: The relationship between the linear velocity v and the angular velocity ***o*** (the Greek letter ***omega)*** for an object with radius ***r*** is v =rω where ω is measured in radians per unit of time and v is distance per unit of time. **r** and ω use the same unit of time and ***r*** and ***v*** use the same linear unit.

Arc length: The length of the curve formed by an arc.

Cosine function: The *x*-value of the point on a unit circle corresponding to a given angle.

Coterminal angles: Explanation of positive and negative angles in a standard position sharing the same terminal side.

Degree: A unit of measure denoting the size of an angle as one-360th of a full revolution of a circle.

Hypotenuse: The side of a right triangle opposite the right angle.

Identities: Statements that are true for all values of the input by which they are defined.

Initial side: The side of an angle from which a rotation begins.

Linear speed: The distance travelled along a straight path by a rotating object within a unit of time; determined by the arc length.

Opposite side: In a right triangle, the side most distant from a given angle.

Period: The smallest interval *P* of a repeating function *f* such that $f(x+P)=f(x)$.

Conversion from degree to radian and vice versa:

$$1^0 = \frac{\pi}{180}\ radian$$

$$1\ radian = \frac{180}{\pi}$$

This formula is used in different types of practical problems.

When we are using this formula, we have to remember two main points.

1. θ should be in radian.
2. l and r should be of same units.

Area of sector: $A = \frac{1}{2} r^2 \theta$.

Angle in standard position: An angle is said to be in standard position if its vertes lies at the origin of a rectangular coordinate system and its initial side along the positive x-axis.

Coterminal angle

Multiple angles may have the same initial and terminal sides. These are called coterminal angles. In general, if angle θ is in degress, then $\theta + 360k$, where $k \in Z$, is an angle coterminal with θ. If angle θ is in radians, then $\theta + 2k\pi$ where $k \in Z$, is an angle coterminal with θ.

Hence, we can say that general angle is $\theta + 2k\pi,\ k \in \mathbf{Z}$.

Concept of quadrants

Each of the four quarters of a circle are called quadrants. When we divide a circle into four equal parts, we get four quadrants.

In the 1st quadrant, the angle will be in the range of $0 \leq \theta \leq 90$

In the 2nd quadrant, the angle will be in the range of $90^\circ \leq \theta \leq 180^\circ$

In the 3rd quadrant, the angle will be in the range of $180^\circ \leq \theta \leq 270^\circ$

In the 4th quadrant, the angle will be in the range of $270^\circ \leq \theta \leq 360^\circ$

Trigonometric ratios:

$sin\, A = \frac{opp}{hyp} = \frac{a}{b}$; $\quad cosec\, A = \frac{hyp}{opp} = \frac{b}{a}$

$cos\, A = \frac{adj}{hyp} = \frac{c}{b}$ $\quad sec\, A = \frac{hyp}{adj} = \frac{b}{c}$

$Tan\, A = \frac{opp}{adj} = \frac{a}{c}$ $\quad \cot A = \frac{adj}{opp} = \frac{c}{a}$

Identity: An equation that is true for all values of the variable for which the terms of the variable are defined.

Fundamental identities:

i. $\sin^2\theta + \cos^2\theta = 1$

ii. $1 + \tan^2\theta = \sec^2\theta$

iii. $1 + \cot^2\theta = cosec^2\theta$

Signs of trigonometric functions in quadrants

If θ lies in quadrant I, then all trigonometric functions are positive.

If θ lies in quadrant II; $\sin\theta$ (positive), $\cos\theta$ (negative) and $\tan\theta$ (negative).

If θ lies in quadrant III; $\sin\theta$ (negative), $\cos\theta$ (negative) and $\tan\theta$ (positive).

If θ lies in quadrant IV; $\sin\theta$ (negative), $\cos\theta$ (positive) and $\tan\theta$ (negative)

CHAPTER 2

FUNCTION VALUES OF SPECIAL ANGLES

INTRODUCTION

In this chapter, we will discuss numerous special angles values. When we start the topic of solving a right triangle using Pythagoras theorem, we frequently work with several angles, such as 45° and 90°. When the terminal line lies on the $x-$axis or the $y-$axis, the angle θ is called a quadrantal angle. We shall determine the values of the trigonometric functions of the quadrantal angles $0^\circ, 90^\circ, 180^\circ, 270^\circ$ and 360° in this chapter.

The right triangle is an imperative geometric stature that has many applications in the world around us. As we begin our study of trigonometry, we will bring to mind what we already know about right triangles.

A right triangle has one right angle and two complementary acute angles. The side opposite the right angle is the longest side of the triangle, known as the hypotenuse. Each side opposite of the acute angle is called a leg.

VALUES OF TRIGONOMETRIC FUNCTIONS FOR $30^\circ, 45^\circ$ and 60°

The relationships among the lengths of the sides of an isosceles right triangle or of the right triangles formed by the altitude to a side of an equilateral triangle make it possible for us to find exact values of the trig functions of the angles of 30°, 45°, and their multiples.

Before calculators and computers were readily available, handbooks that contained tables of values were a common tool of mathematicians. Now calculators or computers will return these values. How were the values in these tables or those displayed by a calculator determined?

Consider a right triangle ABC with $m<C=90^\circ$ and sides a,b,c as shown in Figure 1.

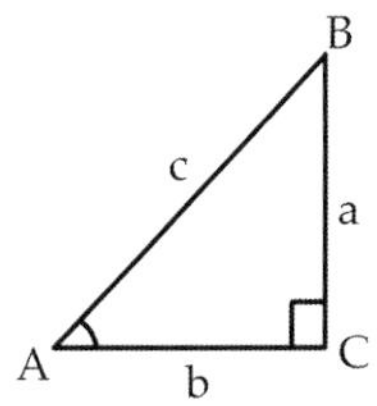

Figure 1

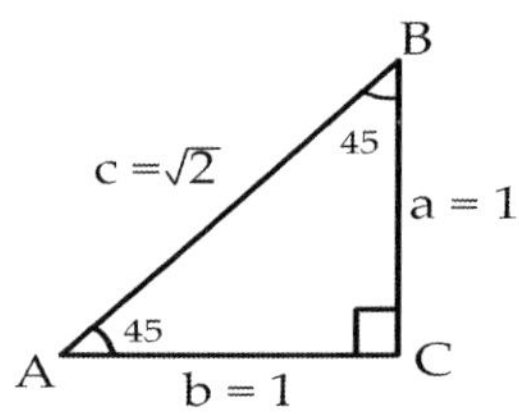

Figure 2

a) Case 1: When $m\angle A=45^0=\frac{\pi}{4}$ **radian**

Then $m\angle B=45^0$

$\Rightarrow \Delta ABC$ is a right isosceles triangle.

As the values of a trigonometric function depend only on the angle and not on the size of the triangle, we can take $a=b=1$

By Pythagoras theorem,

$c^2=a^2+b^2$

$\Rightarrow c^2=1+1=2$

$c=\sqrt{2}$

$\therefore$ Using the triangle with $a=b=1$ and $c=\sqrt{2}$:

$sin\,45^\circ = \frac{a}{c} = \frac{1}{\sqrt{2}}$; $\csc 45^\circ = \frac{1}{\sin 45^\circ} = \sqrt{2}$

$\cos 45^\circ = \frac{b}{c} = \frac{1}{\sqrt{2}}$; $\sec 45^\circ = \frac{1}{\cos 45^\circ} = \sqrt{2}$

$\tan 45^\circ = \frac{a}{b} = 1$; $\cot 45^\circ = \frac{1}{\tan 45^\circ} = 1$

b) Case 2: When $m\angle A = 30^\circ = \frac{\pi}{6}$ radian

Then $m\angle B = 60^\circ$

By elementary geometry, in a right triangle, the measure of the side opposite to 30° is half of the hypotenuse.

Let $c = 2,$ then $a = 1$

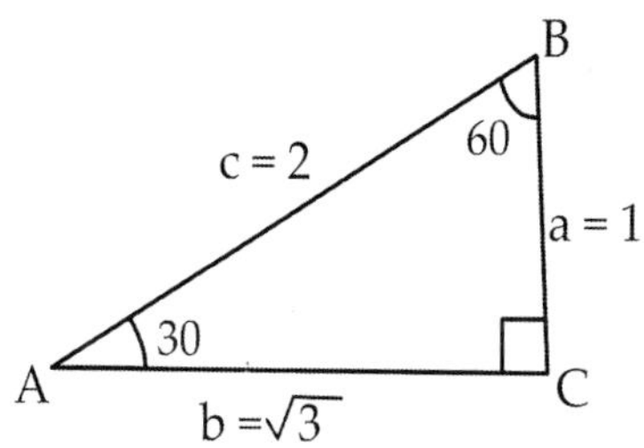

Figure 3

$\therefore$ By Pythagoras theorem, $a^2 + b^2 = c^2$

$\Rightarrow b^2 = c^2 - a^2$

$b^2 = 4 - 1 = 3$

$b = \sqrt{3}.$

$\therefore$ Using the triangle with $a=b=\sqrt{3}$ and $c = 2$:

$sin\ 30^\circ = \frac{a}{c} = \frac{1}{2}$; $\csc 30^\circ = \frac{1}{\sin 45^\circ} = \sqrt{2}$

$\cos 30^\circ = \frac{b}{c} = \frac{\sqrt{3}}{2}$; $\sec 30^\circ = \frac{1}{\cos 45^\circ} = \frac{2}{\sqrt{3}}$

$\tan 30^\circ = \frac{a}{b} = \frac{1}{\sqrt{3}}$;

$\cot 30^\circ = \frac{1}{\tan 30^\circ} = \sqrt{3}$;

c) Case 3: When $m\angle A = 60^\circ = \frac{\pi}{3}$ radian

Then $m\angle B = 30^\circ$.

By elementary geometry, in a right triangle the measure of the side opposite to 30° is half of the hypotenuse.

- Let $c = 2$ then $b = 1$

$\therefore$ By Pythagoras theorem, $a^2 + b^2 = c^2$

$\Rightarrow a^2 = c^2 - b^2$

$a^2 = 4 - 1 = 3$

$a = \sqrt{3}$.

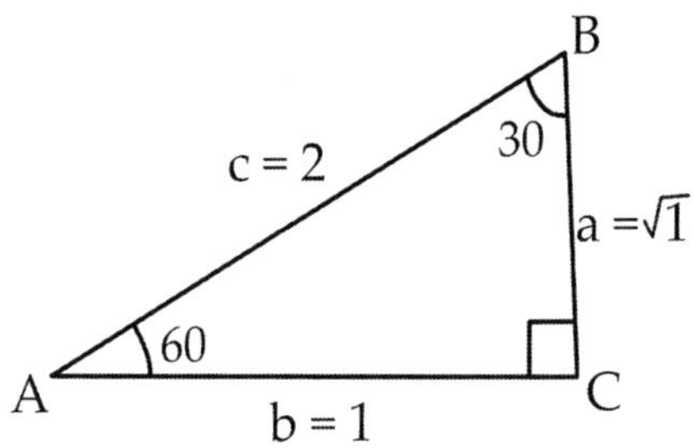

Figure 4

$\therefore$ Using the triangle with $a = \sqrt{3}$, $b = 1$ and $c = 2$:

$\sin 60^\circ = \frac{a}{c} = \frac{\sqrt{3}}{2}$; $\csc 60^\circ = \frac{1}{\sin 60^\circ} = \frac{2}{\sqrt{3}}$

$\cos 60^\circ = \frac{b}{c} = \frac{1}{2}$; $\sec 60^\circ = \frac{1}{\cos 60^\circ} = 2$

$\tan 60^\circ = \frac{a}{b} = \sqrt{3}$; $\cot 60^\circ = \frac{1}{\tan 60^\circ} = \frac{1}{\sqrt{3}}$.

DETERMINING REFERENCE ANGLES

We have discussed finding the sine and cosine for angles in the first quadrant, but what if our angle is in another quadrant? For any given angle in the first quadrant, there is an angle in the second quadrant with the same sine value. Because the sine value is the y-coordinate on the unit circle, the other angle with the same sine will share the same y-value, but have the opposite x-value. Therefore, its cosine value will be the opposite of the first angle's cosine value.

Similarly, there will be an angle in the fourth quadrant with the same cosine as the original. The angle with the same cosine will share the same x-value but will have the opposite y-value. Therefore, its sine value will be the opposite of the original angle's sine value.

$\theta + 2k\pi = \theta$, where $k \in Z$.

Given an angle between 0 and 2π, find its reference angle.

1. An angle in the first quadrant is its own reference angle.
2. If an angle lies in the second quadrant, then its reference angle is $\pi - \theta$.
3. If an angle lies in the third quadrant, then its reference angle is $\pi + \theta$ or $-\pi + \theta$.
4. If an angle lies in the fourth quadrant, then its reference angle is $2\pi - \theta$.

ANGLES WITH A GIVEN FUNCTION VALUE

Since coterminal angles have the same value for a function, there are an unlimited number of angles that have the same value for a trigonometric function. Even when we restrict the angles to the interval of $0° - 360°$, there are usually two angles that have the same function value. All the angles that have the same function value also have the same reference angle. The quadrants for the angle are determined by the sign of the function value. The above relationships are used to find the angle θ, once the reference angle is found from a table or a calculator.

Example. Find values of all the trigonometric functions of the following angles

i. 420° **ii.** 390° **iii.** 765° **iv.** $-\frac{7\pi}{4}$

Solution. We know that $\theta + 2k\pi = \theta$, where $k \in Z$.,

i. We can write $420^\circ = 60^\circ + 1(360^\circ)$ $(k=1)$

So, by above formula the angle will be equal to $\theta = 60^0$.

Now, it is important to understand that 60° and 420° will have the same values.

Hence, we can determine other trigonometric values, such as:

$\sin 420^\circ = \sin 60^\circ = \frac{\sqrt{3}}{2};$ $\csc 420^\circ = \frac{2}{\sqrt{3}};$

$\cos 420^\circ = \cos 60^\circ = \frac{1}{2};$ $\sec 420^\circ = 2;$

$\tan 420^\circ = \tan 60^\circ = \sqrt{3};$ $\cot 420^\circ = \frac{1}{\sqrt{3}};$

we know that $\theta + 2k\pi = \theta$, where $k \in Z$.·

ii. We can write $390^\circ = 30^\circ + 1(360^\circ)$ $(k=1)$

Hence, by the above formula, the angle will be equal to $\theta = 30^\circ$.

Now, it is important to understand that 30° and 390° will have the same values.

Hence, we can determine other trigonometric values, such as:

$\sin 390^\circ = \sin 30^\circ = \frac{1}{2};$ $\csc 390^\circ = 2;$

$\cos 390^\circ = \cos 30^\circ = \frac{\sqrt{3}}{2};$ $\sec 390^\circ = \frac{2}{\sqrt{3}};$

$\tan 390^\circ = \tan 30^\circ = \frac{1}{\sqrt{3}};$ $\cot 390^\circ = \sqrt{3};$

iii. We can write $765^\circ = 45^\circ + 2(360^0)$ $(k=2)$

Hence, by the above formula, the angle will be equal to $\theta = 45°$.

Now, it is important to understand that $45°$ and $765°$ will have the same values.

Hence, we can determine other trigonometric values, such as:

$\sin 765° = \sin 45° = \frac{1}{\sqrt{2}}$; $\csc 765° = \sqrt{2}$;

$\cos 765° = \cos 45° = \frac{1}{\sqrt{2}}$; $\sec 765° = \sqrt{2}$;

$\tan 765° = \tan 45° = \sqrt{3}$; $\cot 765° = 1$;

iv. We can write $-\frac{7\pi}{4} = \frac{\pi}{4} + (-1)2\pi$ $(k = -1)$

Hence, by the above formula, the angle will be equal to $\theta = \frac{\pi}{4}$.

Now, it is important to understand that $\frac{\pi}{4}$, and $\frac{7\pi}{4}$ will have the same values.

Hence, we can determine other trigonometric values, such as:

$\sin\left(\frac{-7\pi}{4}\right) = \sin\frac{\pi}{4} = \frac{\sqrt{3}}{2}$; $\csc\left(-\frac{7\pi}{4}\right) = \frac{2}{\sqrt{3}}$;

$\cos\left(-\frac{7\pi}{4}\right) = \cos\frac{\pi}{4} = \frac{1}{2}$; $\sec\left(-\frac{7\pi}{4}\right) = 2$;

$\tan\left(-\frac{7\pi}{4}\right) = \tan\frac{\pi}{4} = \sqrt{3}$; $\cot\left(-\frac{7\pi}{4}\right) = \frac{1}{\sqrt{3}}$;

Example. Prove that:

i. $\sin\frac{\pi}{3}\cos\frac{\pi}{6} + \cos\frac{\pi}{3}\sin\frac{\pi}{6} = \sin\frac{\pi}{2}$.

ii. $4\tan 60°\ \tan 30°\ \tan 45°\ \sin 30°\ \cos 60° = 1$.

iii. $2\sin 45° + \frac{1}{2}\text{cosec}45° = \frac{3}{\sqrt{2}}$.

iv. $\sin^2\frac{\pi}{6} + \sin^2\frac{\pi}{3} + \tan^2\frac{\pi}{4} = 2$.

v. $\sin^2\frac{\pi}{6} : \sin^2\frac{\pi}{4} : \sin^2\frac{\pi}{3} : \sin^2\frac{\pi}{2} = 1:2:3:4$.

Solution:

$$\sin\frac{\pi}{3}\cos\frac{\pi}{6}+\cos\frac{\pi}{3}\sin\frac{\pi}{6}=\sin\frac{\pi}{2}.$$

Consider the L. H. S:

$$\sin\frac{\pi}{3}\cos\frac{\pi}{6}+\cos\frac{\pi}{3}\sin\frac{\pi}{6}$$

$$=\frac{\sqrt{3}}{2}\times\frac{\sqrt{3}}{2}=\frac{1}{2}\times\frac{1}{2}$$

$$=\frac{3}{4}+\frac{1}{4}=1=\text{RHS.}$$

$$4\tan 60^\circ \tan 30^\circ \tan 45^\circ \sin 30^\circ \cos 60^\circ = 1$$

Consider the L. H. S:

$$= 4\tan 60^\circ \tan 30^\circ \tan 45^\circ \sin 30^\circ \cos 60^\circ$$

$$=4.\sqrt{3}\,.\,\frac{1}{\sqrt{3}}.1.\,\frac{1}{2}.\,\frac{1}{2}$$

$$=1.$$

$$2\sin 45^\circ+\frac{1}{2}\operatorname{cosec}45^\circ=\frac{3}{\sqrt{2}}$$

Consider the L. H. S:

$$2\sin 45^\circ+\frac{1}{2}\operatorname{cosec}45^\circ$$

$$=2\,.\,\frac{1}{\sqrt{2}}+\frac{1}{2}.\sqrt{2}$$

$$=\frac{4+2}{2\sqrt{2}}=\frac{6}{2\sqrt{2}}=\frac{3}{\sqrt{2}}.$$

$$\sin^2\frac{\pi}{6}+\sin^2\frac{\pi}{3}+\tan^2\frac{\pi}{4}=2$$

Consider the L.H.S:

$$=\left(\frac{1}{2}\right)^2+\left(\frac{\sqrt{3}}{2}\right)^2+(1)^2$$

$$=\frac{1}{4}+\frac{3}{4}+1=\frac{1+3+4}{4}=\frac{8}{4}=2.$$

$$\sin^2\frac{\pi}{6}:\sin^2\frac{\pi}{4}:\sin^2\frac{\pi}{3}:\sin^2\frac{\pi}{2}=1:2:3:4$$

Consider the L.H.S:

$$=\left(\frac{1}{2}\right)^2:\left(\frac{1}{\sqrt{2}}\right)^2:\left(\frac{\sqrt{3}}{2}\right)^2:(1)^2$$

$$=\frac{1}{4}:\frac{1}{2}:\frac{3}{4}:1;\ \text{Multiply by4}$$

$=1:2:3:4.$

THE VALUES OF THE TRIGONOMETRIC FUNCTIONS OF QUADRANTAL ANGLES

When the terminal line lies on the $x-$axis or the $y-$axis, the angle θ is called a quadrantal angle. Now we shall find the values of trigonometric functions of the quadrantal angles of $0°, 90°, 180°, 270°, 360°$.

When $\theta = 0°$:

The point $(1,0)$ lies on the terminal side of angle $0°$.

$\Rightarrow \quad x = 1 \text{ and } y = 0.$

Hence, $r = \sqrt{x^2 + y^2} = 1$

$\therefore \sin 0° = \frac{y}{r} = \frac{0}{1} = 0$; $\quad \csc 0° = \frac{1}{\sin 0°} = \frac{1}{0}(undefined)$;

$\cos 0° = \frac{x}{r} = \frac{1}{1} = 1$; $\quad \sec 0° = \frac{1}{\cos 0°} = 1$;

$\tan 0° = \frac{y}{x} = \frac{0}{1} = 0$; $\quad \cot 0° = \frac{1}{\tan 0°} = \frac{1}{0}(undefined)$;

When $\theta = 90°$

The point $(0,1)$ lies on the terminal side of angle $90°$.

$\Rightarrow \quad x = 0 \text{ and } y = 1.$

Hence, $r = \sqrt{x^2 + y^2} = 1$

$\therefore \sin 90° = \frac{y}{r} = \frac{1}{1} = 1$; $\quad \csc 90° = \frac{1}{\sin 90°} = 1$;

$\cos 90° = \frac{x}{r} = \frac{0}{1} = 0$; $\quad \sec 90° = \frac{1}{\cos 90°} = \frac{1}{0}(undefined)$;

$\tan 90° = \frac{y}{x} = \frac{1}{0}(undefined)$; $\quad \cot 90° = \frac{1}{\tan 90°} = 0$;

When $\theta = 180°$

The point $(-1,0)$ lies on the terminal side of angle $180°$.

$\Rightarrow \quad x = -1 \text{ and } y = 0.$

Hence, $r = \sqrt{x^2 + y^2} = 1$

$\therefore \sin 180° = \frac{y}{r} = \frac{0}{1} = 0;$ $\csc 180° = \frac{1}{\sin 180°} = \frac{1}{0}(undefined);$

$\cos 180° = \frac{x}{r} = \frac{-1}{1} = -1;$ $\sec 180° = \frac{1}{\cos 180°} = -1;$

$\tan 180° = \frac{y}{x} = \frac{0}{-1} = 0;$ $\cot 180° = \frac{1}{\tan 180°} = \frac{1}{0}(undefined);$

When $\theta = 270°$.

The point $(0,-1)$ lies on the terminal side of angle $270°$.

$\Rightarrow$ $x = 0$ and $y = -1$,

Hence, $r = \sqrt{x^2 + y^2} = 1$

$\therefore \sin 270° = \frac{y}{r} = \frac{-1}{1} = -1;$ $\csc 270° = \frac{1}{\sin 270°} = -1;$

$\cos 270° = \frac{x}{r} = \frac{0}{1} = 0;$ $\sec 270° = \frac{1}{\cos 270°} = \frac{1}{0}(undefined);$

$\tan 270° = \frac{y}{x} = \frac{-1}{0}(undefined);$ $\cot 270° = \frac{1}{\tan 270°} = 0;$

DOMAINS OF TRIGONOMETRIC FUNCTIONS AND TRIGONOMETRIC IDENTITIES

Here are the list of the trigonometric functions and fundamental identities presented so far, with their domains and ranges.

$\sin\theta$, *for all* $\theta \in \mathbb{R}$.

$\cos\theta$, *for all* $\theta \in \mathbb{R}$.

$cosec\,\theta = \frac{1}{\sin\theta}$, *for all* $\theta \in \mathbb{R}$, *but* $\theta \neq n\pi, n \in \mathbb{Z}$.

$sec\,\theta = \frac{1}{\text{Cos } \theta}$, *for all* $\theta \in \mathbb{R}$, *but* $\theta \neq \left(\frac{2n+1}{2}\right)\pi, n \in \mathbb{Z}$.

$\tan\theta = \frac{\sin\theta}{\cos\theta}$, *for all* $\theta \in \mathbb{R}$, *but* $\theta \neq (2n+1)\frac{\pi}{2}$, $n \in \mathbb{Z}$.

$\cot\theta = \frac{\cos\theta}{\sin\theta}$, *for all* $\theta \in \mathbb{R}$, *but* $\theta \neq n\pi$, $n \in \mathbb{Z}$.

$\sin^2\theta + \cos^2\theta = 1$, *for all* $\theta \in \mathbb{R}$.

$1+\tan^2\theta=\sec^2\theta$, $for\ all\ \theta\in\mathbb{R}, but\ \theta\neq(2n+1)\frac{\pi}{2},\ n\in\mathbb{Z}.$

$1+\cot^2\theta=cosec^2\theta$, $for\ all\ \theta\in\mathbb{R}, but\ \theta\neq n\pi,\ n\in\mathbb{Z}.$

DOMAIN AND RANGE TABLE FOR TRIGONOMETRIC FUNCTIONS

Function	Domain (n is an integer)	Range
Sine	All real numbers	$[-1,1]$
Cosine	All real numbers	$[-1,1]$
Tangent	All real numbers except $\frac{\pi}{2}+n\pi$	All real numbers
Cotangent	All real numbers except $n\pi$	All real numbers
Secant	All real numbers except $\frac{\pi}{2}+n\pi$	$(-\infty,-1]\cup[1,\infty)$
Cosecant	All real numbers except $n\pi$	$(-\infty,-1]\cup[1,\infty)$

TIPS FOR PROVING AN IDENTITY

To prove an identity, use the following tips

- Work with the complicated side.
- Use basic identities to rewrite some functions.
- Remove parenthesis.
- Find the common denominator to add fractions.
- Simplify complex fractions and reduce fractions to lowest form.

APPLICATION OF TRIGONOMETRIC FUNCTIONS AND TRIGONOMETRIC IDENTITIES

Problem 1. Prove that $\cos^4\theta-\sin^4\theta=\cos^2\theta-\sin^2\theta$,

Solution. Consider the LHS: $\cos^4\theta-\sin^4\theta$

$$=\left(\cos^2\theta\right)^2-\left(\sin^2\theta\right)^2$$

$$=\left(\cos^2\theta+\sin^2\theta\right)\left(\cos^2\theta-\sin^2\theta\right)$$

$$=(1)\left(\cos^2\theta-\sin^2\theta\right)\quad\left(\because\sin^2\theta+\cos^2\theta=1\right)$$

$$=\cos^2\theta-\sin^2\theta=R.H.S.$$

Problem 2. $(cosec\theta-\cot\theta)^2=\dfrac{1-\cos\theta}{1+cos\theta}$

Solution: Consider the LHS: $(cosec\theta-\cot\theta)^2$

$$=\left(\frac{1}{\sin\theta}-\frac{\cos\theta}{\sin\theta}\right)^2$$

$$=\left(\frac{1-\cos\theta}{\sin\theta}\right)^2$$

$$=\frac{(1-\cos\theta)^2}{\sin^2\theta}$$

$$=\frac{(1-\cos\theta)^2}{1-\cos^2\theta}$$

$$=\frac{(1-\cos\theta)(1-\cos\theta)}{(1-\cos\theta)(1+\cos\theta)}$$

$$=\frac{1-\cos\theta}{1+\cos\theta}\quad\text{proved.}$$

Problem 3. $\dfrac{1}{1+\sin\theta}+\dfrac{1}{1-\sin\theta}=2\sec^2\theta$

Solution: Consider the LHS: $\dfrac{1}{1+sin\theta}+\dfrac{1}{1-\sin\theta}$

$$=\frac{1-\sin\theta+1+\sin\theta}{1+sin^2\theta}$$

$$=\frac{2}{\cos^2\theta}$$

$$=2\sec^2\theta\quad\text{proved}.$$

Problem 4. $\sqrt{\dfrac{1-\sin\theta}{1+\sin\theta}} = \sec\theta - \tan\theta$

Solution: Consider the LHS: $\sqrt{\dfrac{1-\sin\theta}{1+\sin\theta}}$

$$= \sqrt{\frac{1-\sin\theta}{1+\sin\theta} \times \frac{1-\sin\theta}{1-\sin\theta}}$$

$$= \sqrt{\frac{(1-\sin\theta)^2}{1-\sin^2\theta}}$$

$$= \frac{\sqrt{(1-\sin\theta)^2}}{\sqrt{\cos^2\theta}}$$

$$= \frac{1-\sin\theta}{\cos\theta}$$

$$= \frac{1}{\cos\theta} - \frac{\sin\theta}{\cos\theta}$$

$$= \sec\theta - \tan\theta.$$

Problem 5. $\sqrt{\dfrac{1+\cos\theta}{1-\cos\theta}} = cosec\theta + \cot\theta$

Solution: Consider the LHS: $\sqrt{\dfrac{1+\cos\theta}{1-\cos\theta}}$

$$= \sqrt{\frac{1+\cos\theta}{1-\cos\theta} \times \frac{1+\cos\theta}{1+\cos\theta}}$$

$$= \sqrt{\frac{(1+\cos\theta)^2}{1-\cos^2\theta}}$$

$$= \frac{\sqrt{(1+\cos\theta)^2}}{\sqrt{\sin^2\theta}}$$

$$= \frac{1+\cos\theta}{\sin\theta}$$

$$= \frac{1}{\sin\theta} + \frac{\cos\theta}{\sin\theta}$$

$= cosec\theta + \cot\theta$.

Problem 6. $\frac{\cot^2\theta - 1}{\cot^2\theta + 1} = 2\cos^2\theta - 1$

Solution: Consider the LHS: $\frac{\cot^2\theta - 1}{\cot^2\theta + 1}$

$$= \frac{\frac{\cos^2\theta}{\sin^2\theta} - 1}{\frac{\cos^2\theta}{\sin^2\theta} + 1}$$

$$= \frac{\frac{\cos^2\theta - \sin^2\theta}{\sin^2\theta}}{\frac{\cos^2\theta + \sin^2\theta}{\sin^2\theta}}$$

$$= \frac{\cos^2\theta - \sin^2\theta}{\cos^2\theta + \sin^2\theta}$$

$$= \cos^2\theta - \sin^2\theta$$

$$= \cos^2\theta - (1 - \cos^2\theta)$$

$$= \cos^2\theta - 1 + \cos^2\theta$$

$= 2\cos^2\theta - 1$ proved.

Problem 7. $\frac{\tan\theta}{1-\cot\theta} + \frac{\cot\theta}{1-\tan\theta} = \sec\theta cosec\theta + 1$

Solution: Consider the RHS: $\frac{\tan\theta}{1-\cot\theta} + \frac{\cot\theta}{1-\tan\theta}$

$$= \frac{\frac{\sin\theta}{\cos\theta}}{1 - \frac{\cos\theta}{\sin\theta}} + \frac{\frac{\cos\theta}{\sin\theta}}{1 - \frac{\sin\theta}{\cos\theta}}$$

$$= \frac{\frac{\sin\theta}{\cos\theta}}{\frac{\sin\theta - \cos\theta}{\sin\theta}} + \frac{\frac{\cos\theta}{\sin\theta}}{\frac{\cos\theta - \sin\theta}{\cos\theta}}$$

$$= \frac{\sin^2\theta}{\cos\theta\ (\sin\theta - \cos\theta)} + \frac{\cos^2\theta}{\sin\theta\ (\cos\theta - \sin\theta)}$$

$$= \frac{\sin^2\theta}{\cos\theta(\sin\theta - cos\theta)} - \frac{\cos^2\theta}{\sin\theta(\sin\theta - \cos\theta)}$$

$$= \frac{\sin^3\theta - \cos^3\theta}{\sin\theta.\ \cos\theta(\sin\theta - \cos\theta)}$$

$$= \frac{(\sin\theta - cos\theta)\ (\cos^2\theta + \sin^2\theta + \sin\theta.\cos\theta)}{\sin\theta.\cos\theta(\sin\theta - \cos\theta)}$$

$$= \frac{\left(\cos^2\theta + \sin^2\theta + \sin\theta.\cos\theta\right)}{\sin\theta.\cos\theta}$$

$$= \frac{\left(1+\sin\theta.\cos\theta\right)}{\sin\theta.\cos\theta}$$

$$= \frac{1}{\sin\theta.\cos\theta} + \frac{\sin\theta.\cos\theta}{\sin\theta.\cos\theta}$$

$= \sec\theta cosec\theta + 1$ proved.

Problem 8. $\frac{\sec\theta - \tan\theta}{\sec\theta + \tan\theta} = 1 - 2\sec\theta\tan\theta + 2\tan^2\theta$

Solution: Consider the LHS: $\frac{\sec\theta - \tan\theta}{\sec\theta + \tan\theta}$

$$= \frac{\sec\theta - \tan\theta}{\sec\theta + \tan\theta} \times \frac{\sec\theta - \tan\theta}{\sec\theta - \tan\theta}$$

$$= \frac{\left(\sec\theta - \tan\theta\right)^2}{\sec^2\theta - \tan^2\theta}$$

$$= \frac{\sec^2\theta + \tan^2\theta - 2\sec\theta\tan\theta}{1 + \tan^2\theta - \tan^2\theta}$$

$$= \frac{1 + \tan^2\theta + \tan^2\theta - 2\sec\theta\tan\theta}{1}$$

$= 1 + 2\tan^2\theta - 2\sec\theta\tan\theta$ proved.

Problem 9. $\frac{1+\tan^2\theta}{1+\cot^2\theta} = \frac{\left(1-\tan\theta\right)^2}{\left(1-\cot\theta\right)^2}$

Solution: Consider the RHS: $\frac{\left(1-\tan\theta\right)^2}{\left(1-cot\theta\right)^2}$

$$= \frac{1+\tan^2\theta - 2\tan\theta}{1+\cot^2\theta - 2\cot\theta}$$

$$= \frac{\sec^2\theta - 2\tan\theta}{cosec^2\theta - 2\cot\theta}$$

$$= \frac{\frac{1}{\cos^2\theta} - 2\frac{\sin\theta}{\cos\theta}}{\frac{1}{\sin^2\theta} - 2\frac{\cos\theta}{\sin\theta}}$$

$$= \frac{\frac{1-2\cos\theta\sin\theta}{\cos^2\theta}}{\frac{1-2\cos\theta\sin\theta}{\sin^2\theta}}$$

$$= \frac{\frac{1}{\cos^2\theta}}{\frac{1}{\sin^2\theta}}$$

$$= \frac{\sec^2\theta}{cosec^2\theta}$$

$$= \frac{1+tan^2\theta}{1+\cot^2\theta} \quad \text{proved.}$$

Problem 10. $(1-\tan\theta)^2 + (1-\cot\theta)^2 = (\sec\theta - cosec\theta)^2$

Solution: Consider the LHS: $(1-\tan\theta)^2 + (1-\cot\theta)^2$

$$= 1+\tan^2\theta - 2\tan\theta + 1 + \cot^2\theta - 2\cot\theta$$

$$= \sec^2\theta - 2\tan\theta + cosec^2\theta - 2\cot\theta$$

$$= \sec^2\theta + cosec^2\theta - 2(\tan\theta + \cot\theta)$$

$$= \sec^2\theta + cosec^2\theta - 2\left(\frac{\cos\theta}{\sin\theta} + \frac{\sin\theta}{\cos\theta}\right)$$

$$= \sec^2\theta + cosec^2\theta - 2\left(\frac{\cos^2\theta + \sin^2\theta}{\cos\theta\sin\theta}\right)$$

$$= \sec^2\theta + cosec^2\theta - 2\left(\frac{1}{\cos\theta\sin\theta}\right)$$

$$= \sec^2\theta + cosec^2\theta - 2\sec\theta\ \text{cosec}\ \theta$$

$$= (\sec\theta - cosec\theta)^2 \quad \text{proved.}$$

Problem 11. $\frac{\cos^3 t - \sin^3 t}{\cos t - \sin t} = 1 + \cos t.\sin t$

Solution: Consider the LHS: $\frac{\cos^3 t - sin^3 t}{\cos t - \sin t}$

$$= \frac{(\cos t - \sin t)\,(\cos^2 t + \sin^2 t + \sin t.\cos t)}{\cos t - sint}$$

$$= (\cos^2 t + \sin^2 t + \sin t.\cos t)$$

$= 1 + \cos t . \sin t$ proved.

Problem 12. $\sec^2 A + \tan^2 A = \left(1 - \sin^4 A\right)\sec^4 A$

Solution: Consider the RHS: $\left(1 - \sin^4 A\right)\sec^4 A$

$= \left(1 - \sin^2 A\right)\left(1 + \sin^2 A\right)\sec^4 A$

$= \cos^2 A\left(1 + \sin^2 A\right)\frac{1}{\cos^4 A}$

$= \frac{1 + \sin^2 A}{\cos^2 A}$

$= \frac{1}{\cos^2 A} + \frac{\sin^2 A}{\cos^2 A}$

$= \sec^2 A + \tan^2 A$ proved.

Problem 13. $\frac{\sec x - \cos x}{1 + \cos x} = \sec x - 1$

Solution: Consider the LHS: $\frac{\sec x \ - \ \cos x}{1 \ + \ \cos x}$

$= \frac{\frac{1}{\cos x} - \cos x}{1 + \cos x}$

$= \frac{\frac{1 - \cos^2 x}{\cos x}}{1 + \cos x}$

$= \frac{1 - \cos^2 x}{\cos x\left(1 + \cos x\right)}$

$= \frac{\left(1 - \cos x\right)\left(1 + \cos x\right)}{\cos x\left(1 + \cos x\right)}$

$= \frac{\left(1 - \cos x\right)}{\cos x}$

$= \frac{1}{\cos x} - \frac{\cos x}{\cos x}$

$= \sec x - 1$ proved.

Problem 14. $\dfrac{\sin x + \cos x}{\tan^2 x - 1} = \dfrac{\cos^2 x}{\sin x - \cos x}$

Solution: $\dfrac{\sin x + \cos x}{\tan^2 x - 1}$

$$= \dfrac{\sin x + \cos x}{\dfrac{\sin^2 x}{\cos^2 x} - 1}$$

$$= \dfrac{\sin x + \cos x}{\dfrac{\sin^2 x - \cos^2 x}{\cos^2 x}}$$

$$= \dfrac{\cos^2 x\,(\sin x + \cos x)}{\sin^2 x - \cos^2 x}$$

$$= \dfrac{\cos^2 x\,(\sin x + \cos x)}{(\sin x - \cos x)(\sin x + \cos x)}$$

$$= \dfrac{\cos^2 x}{(\sin x - \cos x)} \quad \text{proved.}$$

Problem 15. $\dfrac{\tan\theta}{\sec\theta - 1} + \dfrac{\tan\theta}{\sec\theta + 1} = 2cosec\theta$

Solution: Consider the LHS: $\dfrac{\tan\theta}{\sec\theta - 1} + \dfrac{\tan\theta}{\sec\theta + 1}$

$$= \dfrac{\tan\theta(sec\theta + 1) + \tan\theta(\sec\theta - 1)}{(\sec\theta - 1)(sec\theta + 1)}$$

$$= \dfrac{\tan\theta\sec\theta + \tan\theta + tan\theta\sec\theta - \tan\theta}{(\sec^2\theta - 1)}$$

$$= \dfrac{2\,tan\theta\,\sec\theta}{\tan^2\theta}$$

$$= \dfrac{2\,sec\theta}{\tan\theta}$$

$$= 2\dfrac{\dfrac{1}{\cos\theta}}{\dfrac{\sin\theta}{\cos\theta}}$$

$$= \dfrac{2}{\sin\theta}$$

$= 2cosec\theta$ proved.

Problem 16. *If* $m = \tan\theta + \sin\theta$ *and* $n = \tan\theta - \sin\theta$, **then proved that** $m^2 - n^2 = 4\sqrt{mn}$.

Solution: Consider the LHS: $m^2 - n^2$

Replace the values of $m = \tan\theta + \sin\theta$ and $n = \tan\theta - \sin\theta$ in the LHS to obtain:

$$= (\tan\theta + \sin\theta)^2 - (\tan\theta - \sin\theta)^2$$

$$= \tan^2\theta + \sin^2\theta + 2\sin\theta\tan\theta - \tan^2\theta - \sin^2\theta + 2\sin\theta\tan\theta$$

$$= 4\sin\theta\tan\theta. \qquad (1)$$

Now, consider the RHS i.e. $4\sqrt{m\,n}$

Put values of m and n in RHS, we obtain:

$$= 4\sqrt{(\tan\theta + \sin\theta)(\tan\theta - \sin\theta)}$$

$$= 4\sqrt{(\tan^2\theta - \sin^2\theta)}$$

$$= 4\sqrt{\left(\frac{\sin^2\theta}{\cos^2\theta} - \sin^2\theta\right)}$$

$$= 4\sqrt{\left(\frac{\sin^2\theta - \sin^2\theta\cos^2\theta}{\cos^2\theta}\right)}$$

$$= 4\sqrt{\frac{\sin^2\theta}{\cos^2\theta}(1 - \cos^2\theta)}$$

$$= 4\frac{\sin\theta}{\cos\theta}\sqrt{(\sin^2\theta)}$$

$$= 4\sin\theta\tan\theta \qquad (2)$$

From (1) and (2) we can say that $\text{LHS} = \text{RHS}$.

Problem 17. Prove that $\frac{\tan\theta + \sec\theta - 1}{\tan\theta - \sec\theta + 1} = \tan\theta + \sec\theta$

Solution: Consider LHS $\frac{\tan\theta + \sec\theta - 1}{\tan\theta - \sec\theta + 1}$

$$= \frac{\tan\theta + \sec\theta - (\sec^2\theta - \tan^2\theta)}{\tan\theta - \sec\theta + 1}$$

$$= \frac{\tan\theta + \sec\theta - (\sec\theta + \tan\theta)(\sec\theta - tan\theta)}{\tan\theta - \sec\theta + 1}$$

$$= \frac{(\tan\theta + \sec\theta)\,(1 - (\sec\theta - \tan\theta)}{\tan\theta - \sec\theta + 1}$$

$$= \frac{(\tan\theta + \sec\theta)\,(\tan\theta - \sec\theta + 1)}{\tan\theta - \sec\theta + 1}$$

$= \tan\theta + \sec\theta = RHS.$

SUMMARY

In this chapter, we discussed the right-angled triangle, as well as some important angles, such as $30^\circ, 45^\circ$ and 60°, and terminal angles, such as $90^\circ, 180^\circ, 270^\circ$ and 360^0. These are the most common angles that we encounter in trigonometric functions.

In section II we discussed identities, some tips solving problems and proving identities. At the end of this chapter, the student will be able to calculate some basic angles and can also easily solve trigonometric identities. The main topic of this chapter was the reference angle. The concept of the reference angle is used in many mathematical problems. Even some complex problems use this concept to solve equations using complex variables.

Review of important terms and formulas

1. **Pythagoras theorem:** $c^2 = a^2 + b^2$
2. **Domains of trigonometric functions and trigonometric identities**

$\sin\theta$, $for\,all\ \theta \in \mathbb{R}.$

$\cos\theta$, $for\,all\ \theta \in \mathbb{R}.$

$Cosec\,\theta = \frac{1}{\sin\theta}$, $for\,all\ \theta \in \mathbb{R}, but\ \theta \neq n\pi, n \in \mathbb{Z}$

$sec\,\theta = \frac{1}{\text{Cos}\ \theta}$, $for\,all\,\theta \in \mathbb{R}, but\ \theta \neq \left(\frac{2n+1}{2}\right)\pi, n \in \mathbb{Z}$

$\tan\theta = \dfrac{\sin\theta}{\cos\theta}$, *for all* $\theta \in \mathbb{R}$, *but* $\theta \neq (2n+1)\dfrac{\pi}{2}$, $n \in \mathbb{Z}$

$\cot\theta = \dfrac{\cos\theta}{\sin\theta}$, *for all* $\theta \in \mathbb{R}$, *but* $\theta \neq n\pi$, $n \in \mathbb{Z}$

$\sin^2\theta + \cos^2\theta = 1$, *for all* $\theta \in \mathbb{R}$

$1 + \tan^2\theta = \sec^2\theta$, *for all* $\theta \in \mathbb{R}$, *but* $\theta \neq (2n+1)\dfrac{\pi}{2}$, $n \in \mathbb{Z}$

$1 + \cot^2\theta = cosec^2\theta$, *for all* $\theta \in \mathbb{R}$, *but* $\theta \neq n\pi$, $n \in \mathbb{Z}$

$2k\pi + \theta = \theta$

How we can calculate or determine a reference angle

Given an angle between 0 and 2π, find its reference angle.

- An angle in the first quadrant is its own reference angle.
- If an angle lies in the second quadrant, then its reference angle is $\pi - \theta$.
- If an angle lies in the third quadrant, then its reference angle is $\pi + \theta$ or $-\pi + \theta$.
- If an angle lies in the fourth quadrant, then its reference angle is $2\pi - \theta$.

To prove an identity, use the following tips

1. Work with the complicated side.
2. Use basic identities to rewrite some functions.
3. Remove parenthesis.
4. Find common denominators to add fractions.
5. Simplify complex fractions and reduce fractions to their lowest form.

CHAPTER 3

TRIGONOMETRIC IDENTITIES - I

INTRODUCTION

In this chapter, we will discuss the fundamental law of trigonometry before discussing numerous trigonometric identities, including double angle identities, half angle identities and triple angle identities. Identities enable us to simplify complicated expressions. They are the basic tools of trigonometry, applied when solving trigonometric equations, just like factoring, finding common denominators, and using special formulas are the basic tools of solving algebraic equations.

In fact, we use algebraic techniques constantly to simplify trigonometric expressions. Basic properties and formulas of algebra, such as the difference of squares formula and the perfect squares formula, will simplify the work involved with trigonometric expressions and equations. We already know that all of the trigonometric functions are related because they are all defined in terms of the unit circle. Consequently, any trigonometric identity can be written in many ways.

To verify the trigonometric identities, we usually start with the more complicated side of the equation and essentially rewrite the expression until it has been transformed into the same expression as the other side of the equation. Sometimes we have to factor expressions, expand expressions, find common denominators, or use other algebraic strategies to obtain the desired result. In this first section, we will work with the fundamental identities: the Pythagorean identities, the even-odd identities, the reciprocal identities, and the quotient identities. We will begin with the Pythagorean identities, which are equations involving trigonometric functions based on the properties of a right triangle.

For the fundamental law of trigonometry, we should know the distance formula between two points at a plane.

Distance formula

Let $P(x_1, y_1)$ and $Q(x_2, y_2)$ be two points. If d denotes the distance between them, then $d = |PQ| = \sqrt{(x_2 - x_1)^2 + (y_2 - y_1)^2}$ is called the distance formula.

Can we write $\cos(A - B) = \cos A - \cos B$?

No. We will check with an example. Let $A = 90°, B = 30°$.

Consider the LHS:

$$\cos(A - B) = \cos(90° - 30°)$$

$$= \cos 60° = \frac{1}{2}$$

Now consider the RHS:

$$\cos A - \cos B = \cos 90° - \cos 30°$$

$$= 0 - \frac{\sqrt{3}}{2} \neq \frac{1}{2}.$$

Now, in order to write an identity that expresses $\cos(A - B)$ in terms of function values of A and B, we will use the relationship between the unit circle and the sine and cosine of an angle.

FUNDAMENTAL LAW OF TRIGONOMETRY

Statement: Let α and β be any two angles, then $\cos(\alpha - \beta) = \cos\alpha \cos\beta + \sin\alpha \sin\beta$.

Proof. Consider a unit circle with its center at origin O. Let α and β be any two angles in a standard position.

The terminal side of angle α intersects the unit circle at $A(\cos\alpha, \sin\alpha)$ and the terminal side of angle β intersects the unit circle at $B(\cos\beta, \sin\beta)$. Take a point C on the unit circle so that $m\angle AOB = \alpha - \beta$.

Join A, B and C, D.

Now angles α, β and $\alpha - \beta$ are in standard position

$\therefore$ The coordinates of A are $(\cos\alpha, \sin\alpha)$.

The coordinates of B are $(\cos\beta, \sin\beta)$.

The coordinates of C are $(\cos(\alpha-\beta), \sin(\alpha-\beta)$).

And the coordinates of D are $(1, 0)$.

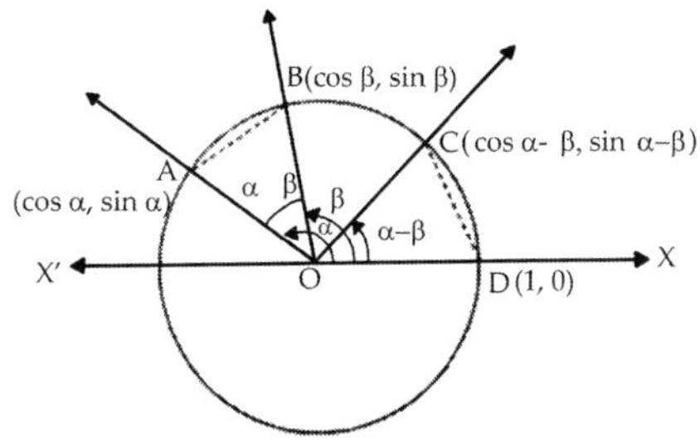

Figure 5

Now $\Delta AOB \, and \, \Delta COD$ are concurrent. (SAS Theorem)

$\therefore \; |AB| = |CD|$

$\Rightarrow |AB|^2 = |CD|^2$

Using the distance formula, we have:

$$(\cos\alpha-\cos\beta)^2+(\sin\alpha-\sin\beta)^2=\left[(\cos(\alpha-\beta)-1\right]^2+\left[(\sin(\alpha-\beta)-0\right]^2$$

$$(\cos^2\alpha+\cos^2\beta-2\cos\alpha\cos\beta)+(\sin^2\alpha+\sin^2\beta-2\sin\alpha\sin\beta)=$$
$$\cos^2(\alpha-\beta)+1-2\cos(\alpha-\beta)+\sin^2(\alpha-\beta)$$

$$(\cos^2\alpha+\sin^2\alpha)+(\cos^2\beta+\sin^2\beta)-2(\sin\alpha\sin\beta+\cos\alpha\cos\beta)=2-\;2\cos(\alpha-\beta)$$

$$1+1-2(\sin\alpha\sin\beta+\cos\alpha\cos\beta)=2-2\cos(\alpha-\beta)$$

$$2-2(\sin\alpha\sin\beta+\cos\alpha\cos\beta)=2-2\cos(\alpha-\beta).$$

Hence $\cos(\alpha-\beta)=\cos\alpha\cos\beta+\sin\alpha\sin\beta$.

At the beginning of this section, we showed that

$\cos(90^\circ-30^\circ)\neq\cos 90^\circ-\cos 30^\circ$.

Does the identity that we proved make it possible to find $\cos(90^\circ-30^\circ)$? Let us check.

$\cos(\alpha-\beta)=\cos\alpha\cos\beta+\sin\alpha\sin\beta$

$\cos(90^\circ-30^\circ)=\cos 90^\circ\cos 30^\circ+\sin 90^\circ\sin 30^\circ$

$\cos 60^\circ = 0.\frac{\sqrt{3}}{2} + 1.\frac{1}{2}$

$\frac{1}{2} = \frac{1}{2}(True)$.

Example. Use $45^\circ - 30^\circ = 15^\circ$ to find the exact value of $\cos 15^\circ$.

Solution. $\text{Cos}(\alpha - \beta) = \cos\alpha\cos\beta + \sin\alpha\sin\beta$

$\cos\left(45^\circ - 30^\circ\right) = \cos 45^\circ \cos 30^\circ + \sin 45^\circ \sin 30^\circ$

$\cos 15^\circ = \frac{1}{\sqrt{2}}.\frac{\sqrt{3}}{2} + \frac{1}{\sqrt{2}}.\frac{1}{2}$

$\cos 15^\circ = \frac{\sqrt{3}}{2\sqrt{2}} + \frac{1}{2\sqrt{2}}$

$\cos 15^\circ = \frac{\sqrt{3}+1.}{2\sqrt{2}}$

DEDUCTIONS FROM THE FUNDAMENTAL LAW

The co-function relationship between cosine and sine can be proved using the fundamental law of trigonometry.

- We know that $\cos(\alpha - \beta) = \cos\alpha\cos\beta + \sin\alpha\sin\beta$.

Replacing α by 90^0 or $\frac{\pi}{2}$, we get

$\cos\,(90^0 - \beta\,) = \cos 90^0 \cos\beta + \sin 90^0 \sin\beta$

$\cos\,(90^0 - \beta\,) = 0.\cos\beta + 1.\,\sin\beta$

$\cos\,(90^0 - \beta\,) = \sin\beta$.

- We know that $\cos\,(\alpha - \beta) = \cos\alpha\cos\beta + \sin\alpha\sin\beta$.

Replacing $\beta = -\frac{\pi}{2}$ we get

$\cos(\alpha - \left(-\frac{\pi}{2}\right)) = \cos\alpha.\cos\left(-\frac{\pi}{2}\right) + \sin\alpha\,\sin\left(-\frac{\pi}{2}\right)$ $\cos\,(\,\alpha + \frac{\pi}{2}\,) = \cos\alpha.0 + \sin\alpha.(-1)$

$\cos(\alpha+\frac{\pi}{2}) = -\sin\alpha$

We know that $\cos(\frac{\pi}{2}-\beta)=\sin\beta$.

Replacing $\beta=\frac{\pi}{2}+\alpha$, we get:

$\cos(\pi/2-\left(\frac{\pi}{2}+\alpha\right))=\sin(\frac{\pi}{2}+\alpha)$

$\cos(-\alpha)=\sin(\frac{\pi}{2}+\alpha)$

$\sin\left(\frac{\pi}{2}+\alpha\right)=\cos\alpha$ $\therefore \cos(-\alpha)=\cos(\alpha)$.

cos(*á* +*â*)

We know that $\cos(\alpha-\beta)=\cos\alpha\cos\beta+\sin\alpha\sin\beta$

Replacing β by $-\beta$ we get

$\cos(\alpha-(-\beta))=\cos\alpha\cos(-\beta)+\sin\alpha\sin(-\beta)$

$\cos(\alpha+\beta)=\cos\alpha\cos\beta-\sin\alpha\sin\beta$ $\therefore \cos(-\beta)=\cos\beta, \sin(-\beta)=-\sin\beta$

$\Rightarrow \cos(\alpha+\beta)=\cos\alpha\cos\beta-\sin\alpha\sin\beta$.

- We know that $\cos(\alpha+\beta)=\cos\alpha\cos\beta-\sin\alpha\sin\beta$.

Replacing α by $\alpha+\frac{\pi}{2}$, we get:

$\cos\left[\left(\frac{\pi}{2}+\alpha\right)+\beta\right]=\cos(\frac{\pi}{2}+\alpha)\cos\beta-\sin(\frac{\pi}{2}+\alpha)\cos\beta$

$\Rightarrow \cos\left[\frac{\pi}{2}+\alpha+\beta\right]=-\sin\alpha\cos\beta-\cos\alpha\sin\beta$

$\Rightarrow -\sin(\alpha+\beta)=-(\sin\alpha\cos\beta+\cos\alpha\sin\beta)$

$\Rightarrow \sin(\alpha+\beta)=\sin\alpha\cos\beta+\cos\alpha\sin\beta$.

We know that $\sin(\alpha+\beta)=\sin\alpha\cos\beta+\cos\alpha\sin\beta$.

Replacing β by $-\beta$, we get

$\sin(\alpha-\beta)=\sin\alpha\cos(-\beta)+\cos\alpha\sin(-\beta)$

$\sin(\alpha-\beta)=\sin\alpha\cos\beta-\cos\alpha\sin\beta$

We know that $\cos(\alpha-\beta)=\cos\alpha\cos\beta+\sin\alpha\sin\beta$.

Replace α with 2π and $\beta = \phi$,

$\therefore \cos(2\pi - \theta) = \cos 2\pi \cos\theta + \sin 2\pi \sin\theta$

$= 1.\cos\theta + 0.\sin\theta$

$= \cos\theta$.

We know that $\sin(\alpha - \beta) = \sin\alpha\cos(-\beta) + \cos\alpha\,\sin(-\beta)$

$\therefore \sin(2\pi - \theta) = \sin 2\pi\cos\theta - \cos 2\pi \sin 0$

$= 0.\cos\theta - 1.\sin\theta$

$= -sin\,\theta$

$$\tan(\alpha + \beta) = \frac{\sin(\alpha+\beta)}{\cos(\alpha+\beta)} = \frac{\sin\alpha\,\cos\beta + \cos\alpha\,\sin\beta}{\cos\alpha\,\cos\beta - \sin\alpha\,\sin\beta}$$

Dividing the numerator and denominator by $\cos\alpha\cos\beta$

$$= \frac{\frac{\sin\alpha\,\cos\beta}{\cos\alpha\cos\beta} + \frac{\cos\alpha\sin\beta}{\cos\alpha\cos\beta}}{\frac{\cos\alpha\cos\beta}{\cos\alpha\cos\beta} - \frac{\sin\alpha\sin\beta}{\cos\alpha\cos\beta}}$$

$$\tan(\alpha + \beta) = \frac{\tan\alpha + \tan\beta}{1 - \tan\alpha\tan\beta}.$$

$$\tan(\alpha - \beta) = \frac{\sin(\alpha-\beta)}{\cos(\alpha-\beta)} = \frac{\sin\alpha\,\cos\beta - \cos\alpha\,\sin\beta}{\cos\alpha\,\cos\beta + \sin\alpha\,\sin\beta}$$

Dividing the numerator and denominator by $\cos\alpha\cos\beta$

$$= \frac{\frac{\sin\alpha\,\cos\beta}{\cos\alpha\cos\beta} - \frac{\cos\alpha\sin\beta}{\cos\alpha\cos\beta}}{\frac{\cos\alpha\cos\beta}{\cos\alpha\cos\beta} + \frac{\sin\alpha\sin\beta}{\cos\alpha\cos\beta}}$$

$$\tan(\alpha - \beta) = \frac{\tan\alpha - \tan\beta}{1 + \tan\alpha\tan\beta}.$$

SOME IMPORTANT RESULTS

- Prove that $\cos(-\theta) = \cos\theta$.

We know from the fundamental law of trigonometry that:

$\cos(\alpha - \beta) = \cos\alpha\cos\beta + \sin\alpha\sin\beta$.

Replacing $\alpha = 0$ and $\beta = \theta$, have:

$\cos(0-\theta) = \cos(0)\cos\theta + \sin(0)\sin\theta$

$\cos(-\theta) = 1.\cos\theta + 0.\sin\theta$

$\cos(-\theta) = \cos\theta$.

Proof that $\sin(-\theta) = -\sin\theta$.

Using the identity $\sin\alpha = \cos(90^\circ - \alpha)$, replace $\alpha = -\theta$

$\sin(-\theta) = \cos(90^\circ - (-\theta))$

$\sin(-\theta) = \cos(90^\circ + \theta)$

$\sin(-\theta) = \cos 90^\circ \cos\theta - \sin 90^\circ \sin\theta$

$\sin(-\theta) = -\sin\theta$.

TRIGONOMETRIC RATIOS OF ALLIED ANGLES

The angles associated with the basic angles of measure θ to a right angle or its multiple are called allied angles. Hence, the angles of measure $90^\circ \pm\theta$, $180^\circ \pm\theta$, $270^\circ \pm\theta$, $360^\circ \pm\theta$, are known as allied angles. Using the fundamental law of trigonometry, $\cos(\alpha-\beta) = \cos\alpha\cos\beta + \sin\alpha\sin\beta$ and its deductions, we derive the following identities:

$\sin(\frac{\pi}{2}-\theta) = \cos\theta$ $\cos(\frac{\pi}{2}-\theta) = \sin\theta$ $\tan(\frac{\pi}{2}-\theta) = \cot\theta$

$\sin(\frac{\pi}{2}+\theta) = \cos\theta$ $\cos(\frac{\pi}{2}+\theta) = -\sin\theta$ $\tan(\frac{\pi}{2}+\theta) = -\cot\theta$

$\sin(\pi-\theta) = \sin\theta$ $\cos(\pi-\theta) = -\cos\theta$ $\tan(\pi-\theta) = -\tan\theta$

$\sin(\pi+\theta) = -\sin\theta$ $\cos(\pi+\theta) = -\cos\theta$ $\tan(\pi+\theta) = \tan\theta$

$\sin(\frac{3\pi}{2}-\theta) = -\cos\theta$ $\cos(\frac{3\pi}{2}-\theta) = -\sin\theta$ $\tan(\frac{3\pi}{2}-\theta) = \cot\theta$

$\sin(\frac{3\pi}{2}+\theta) = -\cos\theta$ $\cos(\frac{3\pi}{2}+\theta) = \sin\theta$ $\tan(\frac{3\pi}{2}+\theta) = -\cot\theta$

$\sin(2\pi-\theta) = -\sin\theta$ $\cos(2\pi-\theta) = \cos\theta$ $\tan(2\pi-\theta) = -\tan\theta$

$\sin(2\pi+\theta) = \sin\theta$ $\cos(2\pi+\theta) = \cos\theta$ $\tan(2\pi+\theta) = \tan\theta$

Tips to remember

The above results also apply to the reciprocal of sine, cosine and tangent. These results are to be applied frequently in the study of trigonometry, and they can be remembered by using the following tips:

- If θ is added to or subtracted from an odd multiple of the right angle, the trigonometric ratios changes into co-ratios and vice versa.

i.e. $sin \rightleftarrows cos$, $\tan \rightleftarrows cot$ $sec \rightleftarrows cosec$ e.g. $\sin(\frac{\pi}{2}-\theta)=\cos\theta$ and $\cos(\frac{3\pi}{2}+\theta)=\sin\theta$.

- If θ is added to or subtracted from an even multiple of $\frac{\pi}{2}$, the trigonometric ratios shall remain the same.
- The sign of the result is determined by the quadrant in which the terminal arm of the angle lies.e.g. $\sin(\pi-\theta)=\sin\theta$, $\tan(\pi+\theta)=\tan\theta$, $\cos(2\pi-\theta)=\cos\theta$.

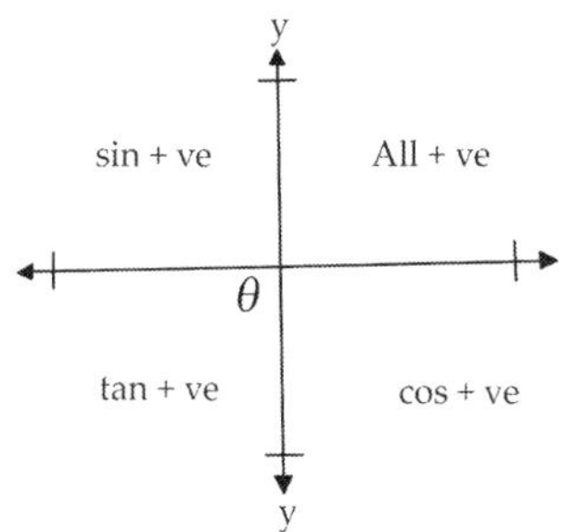

Figure 6

Table:

Measure of the angle	Quadrant
$\frac{\pi}{2}-\theta$	I
$\frac{\pi}{2}+\theta\, or\, \pi-\theta$	II
$\pi+\theta\, or \frac{3\pi}{2}-\theta$	III

$\frac{3\pi}{2}+\theta \, or \, 2\pi-\theta$	IV

Before completing practice questions, revise the important formulas are written below.

1) $\sin(\alpha+\beta)=\sin\alpha\,\cos\beta+\cos\alpha\sin\beta$

2) $\sin(\alpha-\beta)=\sin\alpha\,\cos\beta-\cos\alpha\sin\beta$

3) $\cos(\alpha+\beta)=\cos\alpha\,\cos\beta-\sin\alpha\sin\beta$

4) $\cos(\alpha-\beta)=\cos\alpha\,\cos\beta+\sin\alpha\sin\beta$

5) $\tan(\alpha+\beta)=\frac{\tan\alpha+\tan\beta}{1-\tan\alpha\tan\beta}$

6) $\tan(\alpha-\beta)=\frac{\tan\alpha-\tan\beta}{1+\tan\alpha\tan\beta}$

7) $\sin(-\theta)=-\sin\theta$

8) $\cos(-\theta)=\cos\theta$

9) $\tan(-\theta)=-\tan\theta$

Problem 1. Find the value of

1) $\cos 75°$

Solution: $\cos(45°+30°)=\cos 45°\cos 30°-\sin 45°\sin 30°$

$=\frac{1}{\sqrt{2}}\frac{\sqrt{3}}{2}-\frac{1}{\sqrt{2}}.\frac{1}{2}$

$=\frac{\sqrt{3}}{2\sqrt{2}}-\frac{1}{2\sqrt{2}}=\frac{\sqrt{3}-1}{2\sqrt{2}}.$

2) $\sin 15°$

Solution: $\sin(45°-30°)=\sin 45°\cos 30°-\cos 45°\sin 30°$

$=\frac{1}{\sqrt{2}}.\frac{\sqrt{3}}{2}-\frac{1}{\sqrt{2}}.\frac{1}{2}$

$=\frac{\sqrt{3}}{2\sqrt{2}}-\frac{1}{2\sqrt{2}}=\frac{\sqrt{3}-1}{2\sqrt{2}}.$

3) $\sin 105°$

Solution: $\sin(60° + 45°) = \sin 60° \cos 45° + \cos 60° \sin 30°$

$= \frac{\sqrt{3}}{2}\frac{1}{\sqrt{2}} + \frac{1}{2}\frac{1}{\sqrt{2}}$

$= \frac{\sqrt{3}}{2\sqrt{2}} + \frac{1}{2\sqrt{2}} = \frac{\sqrt{3}+1}{2\sqrt{2}}.$

$\cos 105°$

Solution: $\cos(60° + 45°) = \cos 60° \cos 45° - \sin 60° \sin 45°$

$= \frac{1}{2}.\frac{1}{\sqrt{2}} - \frac{\sqrt{3}}{2}.\frac{1}{\sqrt{2}}$

$= \frac{1}{2\sqrt{2}} - \frac{\sqrt{3}}{2\sqrt{2}} = \frac{1-\sqrt{3}}{2\sqrt{2}}.$

$\tan 105°$

Solution: $\tan(60° + 45°) = \frac{\tan 60° + \tan 45°}{1 - \tan 60°.\tan 45°}$

$= \frac{\sqrt{3}+1}{1-\sqrt{3}\times 1} = \frac{\sqrt{3}+1}{1-\sqrt{3}}.$

Problem 2. Prove that:

$\boldsymbol{sin}(180^0 - \theta) = \boldsymbol{sin}\ \theta$

Solution: LHS = $\sin(180° - \theta)$

$= \sin 180° \cos\theta - \cos 180° \sin\theta$

$= (0)\cos\theta - (-1)\sin\theta$

$= \sin\theta$ = RHS.

$\cos(270° + è) = \sin è$

Solution: LHS = $\sin(270° + \theta)$

$= \cos 270° \cos\theta - \sin 270° \sin\theta$

$= (0)\cos\theta - (-1)\sin\theta$

$= \sin\theta$ =RHS

$\tan(180° + \theta) = \tan\theta$

Solution: LHS = $\tan(180° + \theta)$

$= \dfrac{\tan 180° + \tan\theta}{1 - \tan 180°.\tan\theta} \therefore (\tan 180° = 0)$

$= \dfrac{0 + \tan\theta}{1 - 0 \times \tan\theta}$

$= \tan\theta =$ RHS.

sin (360° – è) = –sin è

Solution: LHS = $\sin(360° - \theta)$

$= \sin 360° \cos\theta - \cos 360° \sin\theta$

$= (0)\cos\theta - (1)\sin\theta$

$= -\sin\theta =$ RHS.

$\cot(360° + \theta) = \cot\theta$

Solution: LHS = $\cot(360° + \theta)$

$= \dfrac{\cos(360° + \theta)}{\sin(360° + \theta)}$

$= \dfrac{\cos 360° \cos\theta - \sin 360° \sin\theta}{\sin 360° \cos\theta + \cos 360° \sin\theta}$

$= \dfrac{(1)\cos\theta - (0)\sin\theta}{(0)\cos\theta + (1)\sin\theta}$

$= \dfrac{\cos\theta}{\sin\theta} = \cot\theta =$ RHS.

$\tan(90° + \theta) = -\cot\theta$

Solution: LHS = $\tan(90° + \theta)$

$= \dfrac{\sin(90° + \theta)}{\cos(90° + \theta)}$

$= \dfrac{\sin 90° \cos\theta + \cos 90° \sin\theta}{\cos 90° \cos\theta - \sin 90° \sin\theta}$

$= \dfrac{(1)\cos\theta + (0)\sin\theta}{(0)\cos\theta - (1)\sin\theta}$

$=\frac{\cos\theta}{-\sin\theta}$

$=-\cot\theta$ = RHS.

Problem 3. Show that

$\sin(\mathbf{x}-\mathbf{y})\cos\mathbf{y}+\cos(\mathbf{x}-\mathbf{y})\sin\mathbf{y}=\sin\mathbf{x}$

Solution: L. H. S = $\sin(x-y)\cos y+\cos(x-y)\sin y$

$=[\sin x\cos y-\cos x\sin y]\cos y+[\cos x\cos y+\sin x\sin y]\sin y$

$=\sin x\cos^2 y-\cos x\sin y\cos y+\cos x\cos y\sin y+\sin x\sin^2 y$

$=\sin x\cos^2 y+\sin x\sin^2 y$

$=\sin x(\cos^2 y+\sin^2 y)=\sin x$ = RHS.

$\cos(\mathbf{x}+\mathbf{y})\cos\mathbf{y}+\sin(\mathbf{x}+\mathbf{y})\sin\mathbf{y}=\cos\mathbf{x}$

Solution: LHS = $\cos(x+y)\cos y+\sin(x+y)\sin y$

$=[\cos x\cos y-\sin x\sin y]\cos y+[\sin x\cos y+\cos x\sin y]\sin y$

$=\cos x\cos^2 y-\sin x\sin y\cos y+\sin x\sin y\cos y+\cos x\sin^2 y$

$=\cos x\cos^2 y+\cos x\sin^2 y$

$=\cos x(\cos^2 y+\sin^2 y)=\cos x$ = RHS.

$\cos(\mathbf{A}+\mathbf{B})\sin(\mathbf{A}-\mathbf{B})=\sin\mathbf{A}\cos\mathbf{A}-\sin\mathbf{B}\cos\mathbf{B}$

Solution: LHS = $\cos(A+B)\sin(A-B)$

$=[\cos A\cos B-\sin A\sin B][\sin A\cos B-\cos A\sin B]$

$=\cos A\cos^2 B\sin A-\sin^2 A\cos B\sin B-\cos B\cos^2 A\sin B+$ $\sin^2 B\cos A\sin A$

$=\cos A\cos^2 B\sin A+\sin^2 B\cos A\sin A-\sin^2 A\cos B\sin B-$

$\cos B\cos^2 A\sin B$

$$= \cos A \sin A\left(\cos^2 B + \sin^2 B\right) - \cos B \sin B\left(\cos^2 A + \sin^2 A\right)$$

$$= \cos A \sin A - \cos B \sin B = \text{RHS}.$$

$$\frac{\tan(\mathbf{x}+\mathbf{y}) - \tan \mathbf{x}}{1+\tan(\mathbf{x}+\mathbf{y})\tan\mathbf{x}} = \frac{\sin \mathbf{x}}{\cos \mathbf{y}}$$

Solution: LHS = $\dfrac{\tan(x+y) - \tan x}{1+\tan(x+y)\tan x}$

$$= \frac{\dfrac{\tan x + \tan y}{1 - \tan x \tan y} - \tan x}{1 + \dfrac{\tan x + \tan y}{1 - \tan x \tan y} \times \tan x}$$

$$= \frac{\dfrac{\tan x + \tan y - \tan x + \tan^2 x \tan y}{1 - \tan x \tan y}}{\dfrac{1 - \tan x \tan y + \tan^2 x + \tan x \tan y}{1 - \tan x \tan y}}$$

$$= \frac{\tan y + \tan^2 x \tan y}{1 + \tan^2 x}$$

$$= \frac{\tan y\left(1 + \tan^2 x\right)}{1 + \tan^2 x}$$

$$= \tan y$$

$$= \frac{\sin y}{\cos y} = \text{RHS}.$$

Problem 4. Suppose that A, B and C are angles of a triangle such that $A+B+C=180^\circ$, prove that $\tan \mathbf{A} + \tan \mathbf{B} + \tan \mathbf{C} = \tan \mathbf{A} \tan \mathbf{B} \tan \mathbf{C}$.

Solution: As we know that:

$$A+B+C=\pi$$

$$A+B=\pi-C$$

$$\Rightarrow \tan(A+B) = \tan(\pi - C)$$

$$\frac{\tan A + \tan B}{1 - \tan A.\tan B} = \frac{\tan \pi - \tan C}{1 + \tan \pi.\tan C}$$

$$\frac{\tan A + \tan B}{1 - \tan A.\tan B} = \frac{\tan 180 - \tan C}{1 + \tan 180.\tan C}$$

$$\frac{\tan A + \tan B}{1 - \tan A.\tan B} = \frac{0 - \tan C}{1 + (0)\tan C}$$

$$\frac{\tan A + \tan B}{1 - \tan A.\tan B} = \frac{-\tan C}{1}$$

$\tan A + \tan B = -\tan C + \tan C \tan A . \tan B$

$\tan A + \tan B + \tan C = \tan A . \tan B . \tan C$ Proved.

Problem 5. Prove that:

$\sin \boldsymbol{\theta} + \cos \boldsymbol{\theta} = \sqrt{2} \sin\left(\boldsymbol{\theta} + \frac{\pi}{4}\right)$

Solution: RHS$=\sqrt{2} \sin\left(\theta + \frac{\pi}{4}\right)$

$= \sqrt{2}\left[\sin\theta \cos 45° + \cos\theta \sin 45°\right]$

$= \sqrt{2}\left[\sin\theta\left(\frac{1}{\sqrt{2}}\right) + \cos\theta\left(\frac{1}{\sqrt{2}}\right)\right]$

$= \left[\sqrt{2}\sin\theta\left(\frac{1}{\sqrt{2}}\right) + \sqrt{2}\cos\theta\left(\frac{1}{\sqrt{2}}\right)\right]$

$= \sin\theta + \cos\theta$ =LHS.

$\sqrt{3}\cos\boldsymbol{\theta} - \sin\boldsymbol{\theta} = 2\cos(\boldsymbol{\theta} + 30°)$

Solution: RHS $= 2\cos(\theta + 30°)$

$= 2\left[\cos\theta\cos 30° - \sin\theta\sin 30°\right]$

$= 2\left[\cos\theta\left(\frac{\sqrt{3}}{2}\right) - \sin\theta\left(\frac{1}{2}\right)\right]$

$= \left[2\cos\theta\left(\frac{\sqrt{3}}{2}\right) \quad 2.\sin\theta\left(\frac{1}{2}\right)\right]$

$= \sqrt{3}\cos\theta - \sin\theta$= LHS.

$\tan(45° - \boldsymbol{\theta}) = \frac{1 - \tan\boldsymbol{\theta}}{1 + \tan\boldsymbol{\theta}}$

Solution: LHS = $\tan(45° + \theta)$

$= \frac{\tan 45° + \tan\theta}{1 - \tan 45° . \tan\theta}$

$= \frac{1 + \tan\theta}{1 - (1)\tan\theta}$

$= \frac{1 + \tan\theta}{1 - \tan\theta}$ = RHS.

$\tan(45° + \boldsymbol{\theta}) = \frac{\cos\theta + \sin\theta}{\cos\theta - \mathbf{sin}\theta}$

Solution: LHS = $\tan(45° + \theta)$

$$= \frac{\sin(45° + \theta)}{\cos(45° + \theta)}$$

$$= \frac{\sin 45° \cos\theta + \cos 45° \sin\theta}{\cos 45° \cos\theta - \sin 45° \sin\theta}$$

$$\frac{\left(\frac{1}{\sqrt{2}}\right)\cos\theta + \left(\frac{1}{\sqrt{2}}\right)\sin\theta}{\left(\frac{1}{\sqrt{2}}\right)\cos\theta - \left(\frac{1}{\sqrt{2}}\right)\sin\theta}$$

$$= \frac{\frac{\cos\theta + \sin\theta}{\sqrt{2}}}{\frac{\cos\theta - \sin\theta}{\sqrt{2}}}$$

$$= \frac{\cos\theta + \sin\theta}{\cos\theta - \sin\theta} = \text{RHS.}$$

$$\frac{\tan(\alpha + \beta)}{\cot(\alpha + \beta)} = \frac{\tan^2\alpha + \tan^2\beta}{1 - \tan^2\alpha . \tan^2\beta}$$

Solution: LHS = $\frac{\tan(\alpha + \beta)}{\cot(\alpha + \beta)}$

$$= \frac{\frac{\tan\alpha + \tan\beta}{1 - \tan\alpha . \tan\beta}}{\frac{1}{\tan(\alpha + \beta)}}$$

$$= \frac{\tan\alpha + \tan\beta}{1 - \tan\alpha . \tan\beta} \times \tan(\alpha + \beta)$$

$$= \frac{\tan\alpha + \tan\beta}{1 - \tan\alpha . \tan\beta} \times \frac{\tan\alpha + \tan\beta}{1 - \tan\alpha . \tan\beta}$$

$$= \frac{\tan^2\alpha + \tan^2\beta}{1 - \tan^2\alpha . \tan^2\beta} = \text{RHS.}$$

$$\cot(\alpha + \beta) = \frac{1 - \cot\alpha . \cot\beta}{\cot\alpha + \cot\beta}$$

Solution:

LHS = $\cot(\alpha + \beta)$

$$= \frac{1}{\tan(\alpha + \beta)}$$

$$= \frac{1}{\frac{\tan\alpha + \tan\beta}{1 - \tan\alpha . \tan\beta}}$$

$$= \frac{1 - \tan\alpha . \tan\beta}{\tan\alpha + \tan\beta}$$

$$= \frac{1 - \left(\frac{1}{\cot\alpha}\right)\left(\frac{1}{\cot\beta}\right)}{\left(\frac{1}{\cot\alpha}\right) + \left(\frac{1}{\cot\beta}\right)}$$

$$= \frac{\frac{\cot\alpha\cot\beta - 1}{\cot\alpha\cot\beta}}{\left(\frac{\cot\beta + \cot\alpha}{\cot\alpha\cot\beta}\right)}$$

$$= \frac{\cot\alpha\cot\beta - 1}{\cot\alpha + \cot\beta} = \text{RHS.}$$

$$\frac{\sin(\alpha+\beta)}{\cos\alpha.\cos\beta} = \tan\alpha + \tan\beta$$

Solution:

$$\text{LHS} = \frac{\sin(\alpha+\beta)}{\cos\alpha.\cos\beta}$$

$$= \frac{\sin\alpha\cos\beta + \cos\alpha\sin\beta}{\cos\alpha.\cos\beta}$$

$$= \frac{\sin\alpha\cos\beta}{\cos\alpha.\cos\beta} + \frac{\cos\alpha\sin\beta}{\cos\alpha.\cos\beta}$$

$$= \frac{\sin\alpha}{\cos\alpha} + \frac{\sin\beta}{\cos\beta}$$

$$= \tan\alpha + \tan\beta = \text{RHS.}$$

$$\frac{\tan(\alpha+\beta)}{\tan(\alpha-\beta)} = \frac{\sin(\alpha+\beta)}{\sin(\alpha-\beta)}$$

Solution:

$$\text{RHS} = \frac{\sin(\alpha+\beta)}{\sin(\alpha-\beta)}$$

$$= \frac{\sin\alpha\cos\beta + \cos\alpha\sin\beta}{\sin\alpha\cos\beta - \cos\alpha\sin\beta}$$

Dividing by $\cos\alpha.\cos\beta$

$$= \frac{\frac{\sin\alpha\cos\beta}{\cos\alpha.\cos\beta} + \frac{\cos\alpha\sin\beta}{\cos\alpha.\cos\beta}}{\frac{\sin\alpha\cos\beta}{\cos\alpha.\cos\beta} - \frac{\cos\alpha\sin\beta}{\cos\alpha.\cos\beta}}$$

$$= \frac{\frac{\sin\alpha}{\cos\alpha} + \frac{\sin\beta}{\cos\beta}}{\frac{\sin\alpha}{\cos\alpha} - \frac{\sin\beta}{\cos\beta}}$$

$$= \frac{\tan(\alpha+\beta)}{\tan(\alpha \quad \beta)} = \text{LHS.}$$

$$\tan\left(x+\frac{\pi}{4}\right) - \tan\left(x-\frac{3\pi}{4}\right) = 0$$

Solution:

$$\text{LHS} = \tan\left(x+\frac{\pi}{4}\right) - \tan\left(x-\frac{3\pi}{4}\right)$$

$$= \tan(x+45°) - \tan(x-135°)$$

$$= \frac{\tan x + \tan 45°}{1 - \tan x.\tan 45°} - \frac{\tan x - \tan 135°}{1 + \tan x.\tan 135°}$$

$$= \frac{\tan x + 1}{1 - \tan x.(1)} - \frac{\tan x - (-1)}{1 + \tan x.(-1)}$$

$$= \frac{\tan x + 1}{1 - \tan x} - \frac{\tan x + 1}{1 - \tan x}$$

$$= 0 \quad = \text{RHS.}$$

$$\cos\left(\frac{\pi}{3}+x\right) - \sin\left(\frac{\pi}{6}-x\right) = 0$$

Solution:

$$\text{LHS} \qquad = \cos\left(\frac{\pi}{3}+x\right) - \sin\left(\frac{\pi}{6}-x\right)$$

$$= \cos(60°+x) - \sin(30°-x)$$

$$= \cos 60° \cos\theta - \sin 60° \sin\theta - \sin 30° \cos x + \cos 30° \sin x$$

$$= \left(\frac{1}{2}\right)\cos\theta - \left(\frac{\sqrt{3}}{2}\right)\sin\theta - \left(\frac{1}{2}\right)\cos x + \left(\frac{\sqrt{3}}{2}\right)\sin x = \quad 0 = \quad \text{RHS.}$$

Problem 6. Prove that:

$$\cos(\alpha+\beta)\cos(\alpha-\beta) = \cos^2\alpha - \cos^2\beta$$

Solution:

$$\text{LHS} = \cos(\alpha+\beta)\cos(\alpha-\beta)$$

$$=[\cos\alpha\cos\beta-\sin\alpha\sin\beta]\,[\cos\alpha\cos\beta+\sin\alpha\sin\beta]$$

$$=\cos^2\alpha\cos^2\beta-\sin^2\alpha\sin^2\beta$$

$$=\cos^2\alpha\left(1-\sin^2\beta\right)-\left(1-\cos^2\alpha\right)\sin^2\beta$$

$$=\cos^2\alpha-\cos^2\alpha\sin^2\beta-\sin^2\beta+\sin^2\beta\cos^2\alpha$$

$$=\cos^2\alpha-\sin^2\beta = \text{RHS.}$$

$\mathbf{\sin(x+y)\sin(x-y)=\sin^2 x-\sin^2 y}$

Solution:

$$\text{LHS} = \sin(x+y)\,\sin(x-y)$$

$$=[\sin x\cos y+\cos x\sin y]\,[\sin x\cos y-\cos x\sin y]$$

$$=\sin^2 x\cos^2 y-\cos^2 x\sin^2 y$$

$$=\sin^2 x\left(1-\sin^2 y\right)-\left(1-\sin^2 x\right)\sin^2 y$$

$$=\sin^2 x-\sin x\sin^2 y-\sin^2 y+\sin^2 y\sin^2 x$$

$$=\sin^2 x-\sin^2 y = \text{RHS Proved.}$$

Problem 7. If $\sin\alpha=\frac{4}{5}$ and $\sin\beta=\frac{12}{13}$, with α and β in the 1st quadrant, find

$\cos\alpha=\sqrt{1-\sin^2\alpha}$; $\cos\beta=\sqrt{1-\sin^2\beta}$

$\cos\alpha=\sqrt{1-\left(\frac{4}{5}\right)^2}$; $\cos\beta=\sqrt{1-\left(\frac{12}{13}\right)^2}$

$\cos\alpha=\sqrt{1-\frac{16}{25}}$; $\cos\beta=\sqrt{1-\frac{144}{169}}$

$\cos\alpha = \sqrt{\frac{9}{25}}$; $\cos\beta = \sqrt{\frac{25}{169}}$

$\cos\alpha = \frac{3}{5}$; $\cos\beta = \frac{5}{13}$

$\sin\alpha = \frac{4}{5}$; $\sin\beta = \frac{12}{13}$

$\sin(\boldsymbol{\alpha}-\boldsymbol{\beta})$

Solution: $\sin(\alpha-\beta) = \sin\alpha\cos\beta - \cos\alpha\sin\beta$

$= \left(\frac{4}{5}\right)\left(\frac{5}{13}\right) - \left(\frac{3}{5}\right)\left(\frac{12}{13}\right) = \frac{4}{13} - \left(\frac{36}{65}\right)$

$= \frac{20-36}{65} = -\frac{16}{65}.$

$\cos(\boldsymbol{\alpha}+\boldsymbol{\beta})$

Solution: $\cos(\alpha+\beta) = \cos\alpha\cos\beta - \sin\alpha\sin\beta$

Now, substitute these values in above formula

$= \left(\frac{3}{5}\right)\left(\frac{5}{13}\right) - \left(\frac{4}{5}\right)\left(\frac{12}{13}\right) = \frac{3}{13} - \left(\frac{48}{65}\right)$

$= \frac{15-48}{65} = -\frac{33}{65}.$

Problem 8. If $\cos\mathbf{A} = \frac{1}{5}$ **and** $\cos\mathbf{B} = \frac{1}{2}$ **, A and B be acute angles. Find the value of:** $\sin A = \sqrt{1-\cos^2 A}$; $\sin B = \sqrt{1-\cos^2 B}$

$\sin A = \sqrt{1-\left(\frac{1}{5}\right)^2}$; $\sin B = \sqrt{1-\left(\frac{1}{2}\right)^2}$

$\sin A = \sqrt{1-\frac{1}{25}}$; $\sin B = \sqrt{1-\frac{1}{4}}$

$\sin A = \sqrt{\frac{24}{25}}$; $\sin B = \sqrt{\frac{3}{4}}$

$\sin A = \frac{\sqrt{24}}{5}$; $\sin B = \frac{\sqrt{3}}{2}$

$\cos A = \frac{1}{5}$; $\cos B = \frac{1}{2}.$

$\sin(\mathbf{A}+\mathbf{B})$

Solution: $\sin(A+B)=\sin A\cos B+\cos A\sin B$

$=\left(\frac{\sqrt{24}}{5}\right)\left(\frac{1}{2}\right)+\left(\frac{1}{5}\right)\left(\frac{\sqrt{3}}{2}\right)=\frac{\sqrt{24}}{10}+\left(\frac{\sqrt{3}}{10}\right)$

$=\frac{\sqrt{24}+\sqrt{3}}{10}$.

$\cos(\mathbf{A}-\mathbf{B})$

Solution: $\cos(A+B)=\cos A\cos B+\sin A\sin B$

$=\left(\frac{1}{5}\right)\left(\frac{1}{2}\right)+\left(\frac{\sqrt{24}}{5}\right)\left(\frac{\sqrt{3}}{2}\right)=\frac{1}{10}+\left(\frac{\sqrt{72}}{10}\right)$

$=\frac{1+\sqrt{72}}{10}=\frac{1+6\sqrt{2}}{10}$.

Problem 9. If $\tan\alpha=\frac{3}{4}$ and $\sec\beta=\frac{13}{5}$ and neither α nor β are in the 1st quadrant, then find $\sin(\alpha+\beta)$.

Solution: Given that neither α nor β are in the 1st quadrant, this implies that **α** belongs to the 3rd quadrant and **β** belongs to the 4th quadrant.

$\sec\beta=\frac{13}{5}$

$\cos\beta=\frac{5}{13}$

$\sec\alpha=\sqrt{1+\tan^2\alpha}$

$\sec\alpha=\sqrt{1+\left(\frac{3}{4}\right)^2}$

$\sec\alpha=\sqrt{1+16}$

$\sec\alpha=\sqrt{\frac{25}{16}}$

$\sec\alpha=\frac{5}{4}$

$\cos\alpha=-\frac{4}{5}$ as α is in third quadrant.

α belongs to the 3rd quadrant and **β** belongs to the 4th quadrant:

$\sin\alpha = \sqrt{1-\cos^2\alpha}$; $\sin\beta = \sqrt{1-\cos^2\beta}$

$\sin\alpha = \sqrt{1-\left(\frac{4}{5}\right)^2}$; $\sin\beta = \sqrt{1-\left(\frac{5}{13}\right)^2}$

$\sin\alpha = \sqrt{1-\frac{16}{25}}$; $\sin\beta = \sqrt{1-\frac{25}{169}}$

$\sin\alpha = \sqrt{\frac{25-16}{25}}$; $\sin\beta = \sqrt{\frac{169-25}{169}}$

$\sin\alpha = \sqrt{\frac{9}{25}}$; $\sin\beta = \sqrt{\frac{144}{169}}$

$\sin\alpha = -\frac{3}{5}$; $\sin\beta = -\frac{12}{13}$

Now $\sin(\alpha+\beta) = \sin\alpha\cos\beta + \cos\alpha\sin\beta$

$= \left(-\frac{3}{5}\right)\left(\frac{5}{13}\right) + \left(-\frac{4}{5}\right)\left(-\frac{12}{13}\right)$

$= \frac{-15}{65} + \left(\frac{48}{65}\right) = \frac{-15+48}{65} = \frac{-33}{65}$.

Problem 10. **Prove that** $\frac{\sin\alpha}{\sec 4\alpha} + \frac{\cos\alpha}{\csc 4\alpha} = \sin 5\alpha$

Solution: LHS= $\frac{\sin\alpha}{\sec 4\alpha} + \frac{\cos\alpha}{\csc 4\alpha}$

$= \frac{\sin\alpha}{\frac{1}{\cos 4\alpha}} + \frac{\cos\alpha}{\frac{1}{\sin 4\alpha}}$

$= \sin\alpha\cos 4\alpha + \cos\alpha\sin 4\alpha$

$= \sin(\alpha+4\alpha) = \sin 5\alpha$ = RHS.

Problem 11. **If** $\tan\beta = \frac{n\sin\alpha\cos\alpha}{1-n\sin^2\alpha}$, **prove that** $\tan(\alpha-\beta) = (1-n)\tan\alpha$

Solution: Given $\tan\beta = \frac{n\sin\alpha\cos\alpha}{1-n\sin^2\alpha}$

LHS= $\tan(\alpha-\beta)$

$= \frac{\tan\alpha + \tan\beta}{1-\tan\alpha\tan\beta}$

$$=\frac{\tan\alpha+\tan\beta}{1-\tan\alpha\tan\beta}$$

$$=\frac{\tan\alpha+\dfrac{n\sin\alpha\cos\alpha}{1-n\sin^2\alpha}}{1-\tan\alpha\dfrac{n\sin\alpha\cos\alpha}{1-n\sin^2\alpha}}$$

$$=\frac{\dfrac{\sin\alpha}{\cos\alpha}+\dfrac{n\sin\alpha\cos\alpha}{1-n\sin^2\alpha}}{1-\left(\dfrac{\sin\alpha}{\cos\alpha}\right)\dfrac{n\sin\alpha\cos\alpha}{1-n\sin^2\alpha}}$$

$$=\frac{\dfrac{\sin\alpha-n\sin^3\alpha+n\sin\alpha\cos^2\alpha}{\cos\alpha\left(1-n\sin^2\alpha\right)}}{\dfrac{\cos\alpha-n\cos\alpha\sin^2\alpha-n\sin^2\alpha\cos\alpha}{\cos\alpha\left(1-n\sin^2\alpha\right)}}$$

$$=\frac{\sin\alpha-n\sin\alpha\left(\sin^2\alpha+\cos^2\alpha\right)}{\cos\alpha}$$

$$=\frac{\sin\alpha-n\sin\alpha}{\cos\alpha}$$

$$=\frac{\sin\alpha(1-n)}{\cos\alpha}=(1-n)\tan\alpha$$ = RHS.

Problem 12. If α, β and γ are the angle of triangle ABC, then prove that:

$$\sin(\alpha+\beta)=\sin\gamma$$

Solution: LHS = $\sin(\alpha+\beta)$

Put $\alpha+\beta=\pi-\gamma$

$$=\sin(\pi-\gamma)$$

$$=\sin\pi\cos\gamma-\cos\pi\sin\gamma$$

$$=(0)\cos\gamma-(-1)\sin\gamma$$

$=\sin\gamma$ = RHS.

$$\cos(\alpha+\beta)=-\cos\gamma$$

Solution: LHS = $\cos(\alpha+\beta)$

Put $\alpha+\beta=\pi-\gamma$

$$=\cos(\pi-\gamma)$$

$= \cos\pi\cos\gamma + \sin\pi\sin\gamma$

$= (-1)\cos\gamma + (0)\sin\gamma$

$= -\cos\gamma$ = RHS.

$\tan(\alpha+\beta)+\tan\gamma = 0$

Solution: As we know that: $\alpha+\beta = \pi-\gamma$

$\tan(\alpha+\beta) = \tan(\pi-\gamma)$

$\tan(\alpha+\beta) = \dfrac{\tan\pi - \tan\gamma}{1+\tan\pi\tan\gamma}$

$\tan(\alpha+\beta) = \dfrac{0-\tan\gamma}{1+(0)\tan\gamma}$

$\tan(\alpha+\beta) = -\tan\gamma$

$\tan(\alpha+\beta)+\tan\gamma = 0$ Proved.

Problem 13. (a) If $\cos\alpha = \frac{1}{7}$ and $\cos\beta = \frac{13}{14}$, then prove that $\alpha - \beta = 60^\circ$, where the terminal rays of α and β are in the 1st quadrant.

$\sin\alpha = \sqrt{1-\cos^2\alpha}$; $\sin\beta = \sqrt{1-\cos^2\beta}$

$\sin\alpha = \sqrt{1-\left(\frac{1}{7}\right)^2}$; $\sin\beta = \sqrt{1-\left(\frac{13}{14}\right)^2}$

$\sin\alpha = \sqrt{1-\frac{1}{49}}$; $\sin\beta = \sqrt{1-\frac{169}{196}}$

$\sin\alpha = \sqrt{\frac{49-1}{49}}$; $\sin\beta = \sqrt{\frac{196-169}{196}}$

$\sin\alpha = \sqrt{\frac{48}{7}}$; $\sin\beta = \sqrt{\frac{27}{196}}$

$\sin\alpha = \frac{4\sqrt{3}}{7}$; $\sin\beta = \frac{3\sqrt{3}}{14}$

$\cos(\alpha-\beta) = \cos\alpha\cos\beta + \sin\alpha\sin\beta$

$= \left(\frac{1}{7}\right)\left(\frac{13}{14}\right) + \left(\frac{4\sqrt{3}}{7}\right)\left(\frac{3\sqrt{3}}{14}\right) = \left(\frac{13}{98}\right) + \left(\frac{12\sqrt{9}}{98}\right)$

$= \left(\frac{13+12*3}{98}\right)$

$= \frac{49}{98} = \frac{1}{2}$

$\cos(\alpha-\beta) = \frac{1}{2}$

$\alpha-\beta = \cos^{-1}\frac{1}{2}$ => $\alpha-\beta = 60^\circ$ proved.

(b) If $\tan\alpha = \frac{5}{6}$ and $\tan\beta = \frac{1}{11}$, then prove that **α+β**=45°, where the terminal rays of a and b are in the 1st quadrant.

Solution: Given that $\tan\alpha = \frac{5}{6}$ and $\tan\beta = \frac{1}{11}$,

we know that: $\tan(\alpha+\beta) = \frac{\tan\alpha - \tan\beta}{1+\tan\alpha\tan\beta}$

$$= \frac{\frac{5}{6} - \frac{1}{11}}{1+\left(\frac{5}{6}\right)\left(\frac{1}{11}\right)} = \frac{\frac{55+6}{66}}{\frac{66-5}{66}}$$

$$= \frac{61}{61} = 1$$

$$\tan(\alpha+\beta) = 1$$

$\alpha+\beta = \tan^{-1}1 \quad = \alpha+\beta = 45°$ proved

SUMMARY

In this chapter, we saw the effectiveness of the Pythagorean Identities along with the Quotient and Reciprocal Identities. Not only did these identities help us compute the values of the circular functions for angles, they were also handy in simplifying expressions involving the circular functions. We also introduced several groups of identities that are very useful in trigonometry problems, particularly for trigonometric equations.

Key terms and formulas

Even/odd functions

$\sin(-\theta) = -\sin\theta, \cos(-\theta) = \cos\theta, \quad \tan(-\theta) = -\tan\theta$

$\csc(-\theta) = -\csc\theta, \sec(-\theta) = \sec\theta, \cot(-\theta) = -\cot\theta$

Fundamental law of trigonometry

Let α and β be any two angles, then $\cos(\alpha-\beta) = \cos\alpha\cos\beta + \sin\alpha\sin\beta$.

Important formulas

$\sin(\alpha+\beta) = \sin\alpha\ \cos\beta + \cos\alpha\sin\beta$

$\sin(\alpha-\beta) = \sin\alpha\ \cos\beta - \cos\alpha\sin\beta$

$\cos(\alpha+\beta) = \cos\alpha\ \cos\beta - \sin\alpha \sin\beta$

$\cos(\alpha-\beta) = \cos\alpha\ \cos\beta + \sin\alpha \sin\beta$

$\tan(\alpha+\beta) = \dfrac{\tan\alpha + \tan\beta}{1-\tan\alpha\tan\beta}$

$\tan(\alpha-\beta) = \dfrac{\tan\alpha - \tan\beta}{1+\tan\alpha\tan\beta}$

Reference angle

Angle	Quadrant
$\frac{\pi}{2}-\theta$	I
$\frac{\pi}{2}+\theta$ or $\pi-\theta$	II
$\pi+\theta$ or $\frac{3\pi}{2}-\theta$	III
$\frac{3\pi}{2}+\theta$ or $2\pi-\theta$	IV

Deductions from the fundamental law

The co-function relationship between cosine and sine can be proved using the fundamental law of trigonometry.

1. We know that $\cos(\alpha-\beta) = \cos\alpha\cos\beta + \sin\alpha\sin\beta$

   Replace α by 90° or $\frac{\pi}{2}$

$\cos(90^0-\beta) = \cos 90^0 \cos\beta + \sin 90^0 \sin\beta$

$\cos(90^0-\beta) = 0.\cos\beta + 1.\ \sin\beta$

$\cos(90^0-\beta) = \sin\beta$

2. We know that $\cos(\alpha-\beta) = \cos\alpha\cos\beta + \sin\alpha\sin\beta$

Replace $\beta = -\frac{\pi}{2}$, we get

$\cos(\alpha - \left(-\frac{\pi}{2}\right)) = \cos\alpha.\cos\left(-\frac{\pi}{2}\right) + \sin\alpha \sin\left(-\frac{\pi}{2}\right) \cos(\alpha+\frac{\pi}{2}) = \cos\alpha.0 + \sin\alpha.(-1)$

$\cos(\alpha+\frac{\pi}{2}) = -\sin\alpha$

3. We know that $\cos\left(\frac{\pi}{2}-\beta\right)=\sin\beta$

Replace $\beta=\frac{\pi}{2}+\alpha$, we get

$\cos\left(\pi/2-\left(\frac{\pi}{2}+\alpha\right)\right)=\sin\left(\frac{\pi}{2}+\alpha\right)$

$\cos(-\alpha)=\sin\left(\frac{\pi}{2}+\alpha\right)$

$\sin\left(\frac{\pi}{2}+\alpha\right)=\cos\alpha$

$\therefore \cos(-\alpha)=\cos(\alpha)\cdot$

$\cos(\alpha+\beta)$

As we know that $\cos(\alpha-\beta)=\cos\alpha\cos\beta+\sin\alpha\sin\beta$

Replacing β by $-\beta$ we get,

$\cos(\alpha-(-\beta))=\cos\alpha\cos(-\beta)+\sin\alpha\sin(-\beta)$

$\cos(\alpha+\beta)=\cos\alpha\cos\beta-\sin\alpha\sin\beta \qquad \therefore \cos(-\beta)=\cos\beta, \sin(-\beta)=-\sin\beta$

$\Rightarrow \cos(\alpha+\beta)=\cos\alpha\cos\beta-\sin\alpha\sin\beta$.

4. We know that $\cos(\alpha+\beta)=\cos\alpha\cos\beta-\sin\alpha\sin\beta$

Replacing α by $\alpha+\frac{\pi}{2}$, we get

$\cos\left[\left(\frac{\pi}{2}+\alpha\right)+\beta\right]=\cos\left(\frac{\pi}{2}+\alpha\right)\cos\beta-\sin\left(\frac{\pi}{2}+\alpha\right)\cos\beta$

$\Rightarrow \quad \cos\left[\frac{\pi}{2}+\alpha+\beta\right]=-\sin\alpha\cos\beta-\cos\alpha\sin\beta$

$\Rightarrow \quad -\sin(\alpha+\beta)=-(\sin\alpha\cos\beta+\cos\alpha\sin\beta)$

$\Rightarrow \quad \sin(\alpha+\beta)=\sin\alpha\cos\beta+\cos\alpha\sin\beta\cdot$

5. We know that $\sin(\alpha+\beta)=\sin\alpha\cos\beta+\cos\alpha\sin\beta$

Replace β by $-\beta$, we get

$\sin(\alpha-\beta)=\sin\alpha\cos(-\beta)+\cos\alpha\sin(-\beta)$

$\sin(\alpha-\beta)=\sin\alpha\cos\beta-\cos\alpha\sin\beta\cdot$

6. We know that $\cos(\alpha-\beta)=\cos\alpha\cos\beta+\sin\alpha\sin\beta$

Replace α with 2π and $\beta=\phi$

$\therefore \cos(2\pi-\theta)=\cos 2\pi\cos\theta+\sin 2\pi\sin\theta$

$=1.\cos\theta+0.\sin\theta$

$=\cos\theta$.

7. We know that $\sin(\alpha-\beta)=\sin\alpha\cos(-\beta)+\cos\alpha\ \sin(-\beta)$

$\therefore \sin(2\pi-\theta)=\sin 2\pi\cos\theta-\cos 2\pi\sin 0$

$=0.\cos\theta-1.\sin\theta$

$=-sin\theta$.

$$\tan(\alpha+\beta)=\frac{\sin(\alpha+\beta)}{\cos(\alpha+\beta)}=\frac{\sin\alpha\ \cos\beta+\cos\alpha\ \sin\beta}{\cos\alpha\ \cos\beta-\sin\alpha\ \sin\beta}$$

Dividing the numerator and denominator by $\cos\alpha\cos\beta$

$$=\frac{\frac{\sin\alpha\ \cos\beta}{\cos\alpha\cos\beta}+\frac{\cos\alpha\sin\beta}{\cos\alpha\cos\beta}}{\frac{\cos\alpha\cos\beta}{\cos\alpha\cos\beta}-\frac{\sin\alpha\sin\beta}{\cos\alpha\cos\beta}}$$

$$\tan(\alpha+\beta)=\frac{\tan\alpha+\tan\beta}{1-\tan\alpha\tan\beta}$$

$$\tan(\alpha-\beta)=\frac{\sin(\alpha-\beta)}{\cos(\alpha-\beta)}=\frac{\sin\alpha\ \cos\beta-\cos\alpha\ \sin\beta}{\cos\alpha\ \cos\beta+\sin\alpha\ \sin\beta}$$

Dividing the numerator and denominator by $\cos\alpha\cos\beta$

$$=\frac{\frac{\sin\alpha\ \cos\beta}{\cos\alpha\cos\beta}-\frac{\cos\alpha\sin\beta}{\cos\alpha\cos\beta}}{\frac{\cos\alpha\cos\beta}{\cos\alpha\cos\beta}+\frac{\sin\alpha\sin\beta}{\cos\alpha\cos\beta}}$$

$$\tan(\alpha-\beta)=\frac{\tan\alpha-\tan\beta}{1+\tan\alpha\tan\beta}.$$

Trigonometric ratios of allied angles

The angles associated with basic angles of measure θ to a right angle or its multiple are called allied angles. Thus, the angles of measure $90^\circ\pm\theta$, $180^\circ\pm\theta,\ 270^\circ\pm\theta,\ 360^\circ\pm\theta$, are known as allied angles.

$\sin(\frac{\pi}{2}-\theta)=\cos\theta$ $\cos(\frac{\pi}{2}-\theta)=\sin\theta$ $\tan(\frac{\pi}{2}-\theta)=\cot\theta$

$\sin(\frac{\pi}{2}+\theta)=\cos\theta$ $\cos(\frac{\pi}{2}+\theta)=-\sin\theta$ $\tan(\frac{\pi}{2}+\theta)=-\cot\theta$

$\sin(\pi-\theta)=\sin\theta$ $\cos(\pi-\theta)=-\cos\theta$ $\tan(\pi-\theta)=-\tan\theta$

$\sin(\pi+\theta)=-\sin\theta$ $\cos(\pi+\theta)=-\cos\theta$ $\tan(\pi+\theta)=\tan\theta$

$\sin(\frac{3\pi}{2}-\theta)=-\cos\theta$ $\cos(\frac{3\pi}{2}-\theta)=-\sin\theta$ $\tan(\frac{3\pi}{2}-\theta)=\cot\theta$

$\sin(\frac{3\pi}{2}+\theta)=-\cos\theta$ $\cos(\frac{3\pi}{2}+\theta)=\sin\theta$ $\tan(\frac{3\pi}{2}+\theta)=-\cot\theta$

$\sin(2\pi-\theta)=-\sin\theta$ $\cos(2\pi-\theta)=\cos\theta$ $\tan(2\pi-\theta)=-\tan\theta$

$\sin(2\pi+\theta)=\sin\theta$ $\cos(2\pi+\theta)=\cos\theta$ $\tan(2\pi+\theta)=\tan\theta$

CHAPTER 4

TRIGONOMETRIC IDENTITIES

INTRODUCTION

In the previous chapter, we saw the importance of the basic trigonometric identities, along with their applications in numerous problems. In this chapter, we will discuss more important identities, such as reduction, double angle identities, half angle identities and triple angle identities. However, before applying these identities, we will present their derivation using the sum/difference formulas of the sine and cosine functions.

Trigonometric identities

An equation involving trigonometric functions that are valid for all values of the angle for which the functions are defined is called a trigonometric identity.

General guidelines for verifying identities

- Know the eight basic relationships and recognize each of their alternative forms.
- Know the procedures for adding and subtracting fractions, reducing fractions, and transforming fractions into equivalent fractions.
- Know factoring and special product techniques.
- Use only substitution and simplification procedures that allow

you to work on exactly one side of an equation.

- Select the side of the equation that appears more complicated and attempt to transform it into the form of the other side of the equation.
- Transform each side of the equation, independently, into the same form.
- Avoid substitutions that introduce radicals.
- Use substitutions to change all trigonometric functions into expressions involving only sine and cosine, and then simplify.
- Multiply the numerator and denominator of a fraction by the conjugate of either.
- Simplify a square root of a fraction using conjugates to transform it into the quotient of perfect squares.

DOUBLE ANGLE IDENTITIES

As we know that

$\sin(\alpha+\beta) = \sin\alpha\cos\beta + \cos\alpha\sin\beta$,

we can replace β with α in the above equation to we obtain

$\sin(\alpha+\alpha) = \sin\alpha\ \cos\alpha + \cos\alpha\ \sin\alpha$

$$\sin 2\alpha = 2\sin\alpha\cos\alpha \qquad \textbf{(4.1)}$$

$\cos(\alpha+\beta) = \cos\alpha\ \cos\beta - \sin\alpha\sin\beta$.

Replace β with α in the above equation to obtain

$\cos(\alpha+\alpha) = \cos\alpha\ \cos\alpha - \sin\alpha\ \sin\alpha$

$$\cos 2\alpha = \cos^2\alpha - \sin^2\alpha \qquad \textbf{(4.2)}$$

Also, we know that

$$\tan(\alpha+\beta) = \frac{\tan\alpha + \tan\beta}{1-\tan\alpha\ \tan\beta}.$$

Replace β with α in above equation to obtain

$$\tan(\alpha+\alpha)=\frac{\tan\alpha+\tan\alpha}{1-\tan\alpha\ \tan\alpha}$$

$$\tan(2\alpha)=\frac{2\ \tan\alpha}{1-\ \tan^2\alpha}. \quad \textbf{(4.3)}$$

Using double angle formulas to verify identities

Problem 1. $\tan\theta=\frac{\sin 2\theta}{1+\cos 2\theta}$

Solution: RHS = $\frac{\sin 2\theta}{1+\cos 2\theta}$

$$=\frac{2\sin\theta\cos\theta}{1+\cos^2\theta-\sin^2\theta}$$

$$=\frac{2\sin\theta\cos\theta}{1+\cos^2\theta-\sin^2\theta}$$

$$=\frac{2\sin\theta\ \cos\theta}{\cos^2\theta+\cos^2\theta}$$

$$=\frac{2\sin\theta\cos\theta}{2\cos^2\theta}$$

$$=\frac{\sin\theta}{\cos\theta}$$

$=\tan\theta=$ LHS.

Problem 2. $\cot\theta=\frac{\sin 2\theta}{1-\cos 2\theta}$

Solution: R. H. S = $\frac{\sin 2\theta}{1-\cos 2\theta}$

$$=\frac{2\sin\theta\ \cos\theta}{1-\cos^2\theta+\sin^2\theta}$$

$$=\frac{2\sin\theta\ \cos\theta}{\sin^2\theta+\cos^2\theta}$$

$$=\frac{2\sin\theta\cos\theta}{2\sin^2\theta}$$

$=\frac{\cos\theta}{\sin\theta}=\cot\theta=$ LHS.

Problem 3. $\dfrac{1+\sin 2A-\cos 2A}{1+\sin 2A+\cos 2A}=\tan A$

Solution: LHS = $\dfrac{1+\sin 2A-\cos 2A}{1+\sin 2A+\cos 2A}$

$$=\frac{1+2\sin A\cos A-\cos^2 A+\sin^2 A}{1+2\sin A\cos A+\cos^2 A-\sin^2 A}$$

$$=\frac{\sin^2 A+2\sin A\cos A+\sin^2 A}{\cos^2 A+2\sin A\cos A+\cos^2 A}$$

$$=\frac{2\sin^2 A+2\sin A\cos A}{2\cos^2 A+2\sin A\cos A}$$

$$=\frac{2\sin A(\sin A+\cos A)}{2\cos A(\cos A+\sin A)}=\frac{\sin A}{\cos A}$$

$$=\tan A=\text{RHS}$$

Problem 4. $\sec 2\mathbf{A}+\tan 2\mathbf{A}=\dfrac{\cos\mathbf{A}+\sin\mathbf{A}}{\cos\mathbf{A}-\mathbf{\sin A}}$

Solution: RHS = $\dfrac{\cos A+\sin A}{\cos A-\sin A}$

$$=\frac{\cos A+\sin A}{\cos A-\sin A}\times\frac{\cos A+\sin A}{\cos A+\sin A}$$

$$=\frac{(\cos A+\sin A)^2}{\cos^2 A-\sin^2 A}$$

$$=\frac{\cos^2 A+\sin^2 A+2\sin A\cos A}{\cos 2A}$$

$$= \frac{1 + \sin 2A}{\cos 2A} = \frac{1}{\cos 2A} + \frac{\sin 2A}{\cos 2A}$$

$$= \sec 2A + \tan 2A = \text{LHS}.$$

Problem 5. $\operatorname{cosec} 2\theta + \cot 2\theta = \cot \theta$

Solution: LHS $= \operatorname{cosec} 2\theta + \cot 2\theta$

$$= \frac{1}{\sin 2\theta} + \frac{\cos 2\theta}{\sin 2\theta}$$

$$= \frac{1 + \cos 2\theta}{\sin 2\theta}$$

$$= \frac{1 + \cos^2 \theta - \sin^2 \theta}{2 \sin\theta \cos\theta}$$

$$= \frac{\cos^2 \theta + \cos^2 \theta}{2 \sin\theta \cos\theta}$$

$$= \frac{2\cos^2 \theta}{2 \sin\theta \cos\theta}$$

$$= \frac{\cos\theta}{\sin\theta} = \cot\theta = \text{RHS}.$$

Note that when using the identities to simplify a trigonometric expression, there are many ways to get the desired results. There is no specific rule to solve or manipulate a side.

Problem 6. $\frac{1 - \cos 2\mathbf{A}}{1 + \mathbf{cos 2A}} = \tan^2 \mathbf{A}$

Solution: LHS $= \frac{1 - \cos 2A}{1 + \cos 2A}$

$$= \frac{1 - \cos^2 A + \sin^2 A}{1 + \cos^2 A - \sin^2 A}$$

$$= \frac{\sin^2 A + \sin^2 A}{\cos^2 A + \cos^2 A}$$

$$= \frac{2\sin^2 A}{2\cos^2 A}$$

$$= \tan^2 A = \text{RHS}.$$

Problem 7. $\frac{\sin 3\theta}{\cos\theta}+\frac{\cos 3\theta}{\sin\theta}=2\cot 2\theta$

Solution: LHS = $\frac{\sin 3\theta}{\cos\theta}+\frac{\cos 3\theta}{\sin\theta}$

$$=\frac{\cos\theta\sin 3\theta+\sin\theta\cos 3\theta}{\cos\theta\sin\theta}$$

$$=\frac{\cos\ (3\theta-\theta)}{\cos\theta\ \sin\theta}$$

$$=\frac{\cos 2\theta}{\cos\theta\sin\theta}$$

$$=\frac{2\ \cos 2\theta}{2\ \cos\theta\sin\theta}$$

$$=\frac{2\ \cos 2\theta}{\sin 2\theta}$$

$$=2\cot 2\theta = \text{RHS.}$$

Problem 8. $\frac{\cot^2\theta-1}{\operatorname{cosec}^2\theta}=\cos 2\theta$

Solution: LHS = $\frac{\cot^2\theta-1}{\operatorname{cosec}^2\theta}$

$$=\frac{\frac{\cos^2\theta}{\sin^2\theta}-1}{\frac{1}{\sin^2\theta}}$$

$$=\frac{\frac{\cos^2\theta-\sin^2\theta}{\sin^2\theta}}{\frac{1}{\sin^2\theta}}$$

$$=\cos^2\theta-\sin^2\theta$$

$$=\cos 2\theta = \text{RHS.}$$

Problem 9. $\cos^4\theta-\sin^4\theta=\frac{1}{\sec 2\theta}$

Solution: LHS = $\cos^4\theta-\sin^4\theta$

$= (\cos^2\theta + \sin^2\theta)(\cos^2\theta - \sin^2\theta)$

$= (1)(\cos 2\theta) = \cos 2\theta$ = RHS.

Problem 10. $\tan 2\theta = \frac{2}{\cot\theta - \tan\theta}$

Solution: Consider the LHS $\tan 2\theta$

$= \frac{2\tan\theta}{1-\tan^2\theta}$

Multiplying and dividing by $\frac{1}{\tan\theta}$, we obtain

$= \frac{2\tan\theta \times \frac{1}{\tan\theta}}{(1-\tan^2\theta)\times\frac{1}{\tan\theta}}$

$= \frac{2}{\frac{1}{\tan\theta} - \frac{\tan^2\theta}{\tan\theta}}$

$= \frac{2}{\cot\theta - \tan\theta}$ = RHS.

REDUCTION FORMULAS AND HALF ANGLE IDENTITIES

The half angle and reduction formulas can be obtained from double angle identities. These formulas are used to reduce the power of a given expression involving even powers of sine and cosine functions. The half angle or reduction formulas play a very important role in higher level studies, particularly in calculus. We can derive these formulas from the double angle identities

Consider $\cos 2\theta = 1 - 2\sin^2\theta$, we can write:

$2\sin^2\theta = (1 - \cos 2\theta)$

$\sin^2\theta = \frac{1-\cos 2\theta}{2}$

$$\sin\theta = \pm\sqrt{\frac{1-\cos 2\theta}{2}} \qquad \textbf{(4.4)}$$

Similarly,

$\cos 2\theta = 2\cos^2\theta - 1$, we can

$2\cos^2\theta = 1 + \cos 2\theta$

$$\cos^2\theta = \frac{1+\cos 2\theta}{2}$$

$$\cos\theta = \pm\sqrt{\frac{1+\cos 2\theta}{2}}. \tag{4.5}$$

For next reduction formula, we can write

$$\tan^2\theta = \frac{\sin^2\theta}{\cos^2\theta}$$

$$= \frac{\frac{1-\cos 2\theta}{2}}{\frac{1+\cos 2\theta}{2}}$$

$$\tan^2\theta = \frac{1-\cos 2\theta}{1+\cos 2\theta}$$

$$\tan\theta = \pm\sqrt{\frac{1-\cos 2\theta}{1+\cos 2\theta}}. \tag{4.6}$$

Using the half angle/reduction formulas to verify identities

Problem 1. If $\cos\theta = \frac{4}{5}$ and the terminal ray of θ are in the first quadrant, find the value of

i. $\sin\frac{\theta}{2}$

Solution: $\sin\frac{\theta}{2} = \sqrt{\frac{1-\cos\theta}{2}} = \sqrt{\frac{1-\frac{4}{5}}{2}} = \sqrt{\frac{\frac{5-4}{5}}{2}} = \sqrt{\frac{1}{5\times 2}} = \frac{1}{\sqrt{10}}.$

ii. $\cos\frac{\theta}{2}$

Solution: $\cos\frac{\theta}{2} = \sqrt{\frac{1+\cos\theta}{2}} = \sqrt{\frac{1+\frac{4}{5}}{2}} = \sqrt{\frac{\frac{5+4}{5}}{2}} = \sqrt{\frac{9}{5\times 2}} = \frac{3}{\sqrt{10}}.$

iii. $\tan\frac{\theta}{2}$

Solution: $\tan\frac{\theta}{2} = \sqrt{\frac{1-\cos\theta}{1+\cos\theta}} = \sqrt{\frac{1-\frac{4}{5}}{1+\frac{4}{5}}} = \sqrt{\frac{\frac{5-4}{5}}{\frac{5+4}{5}}} = \sqrt{\frac{1}{9}} = \frac{1}{3}$

Problem 2. If $\cos\theta = -\frac{5}{13}$ and the terminal sides of θ are in the second quadrant, find the value of

i. $\sin\frac{\theta}{2}$

Solution: $\sin\frac{\theta}{2} = \sqrt{\frac{1-\cos\theta}{2}} = \sqrt{\frac{1-\left(-\frac{5}{13}\right)}{2}} = \sqrt{\frac{\frac{13+5}{13}}{2}} = \sqrt{\frac{18}{13\times 2}} = \sqrt{\frac{9}{13}} = \frac{3}{\sqrt{13}}$.

ii. $\cos\frac{\theta}{2}$

Solution: $\cos\frac{\theta}{2} = \sqrt{\frac{1+\cos\theta}{2}} = \sqrt{\frac{1+\left(\frac{-5}{13}\right)}{2}} = \sqrt{\frac{\frac{13-5}{13}}{2}} = \sqrt{\frac{8}{13\times 2}} = \sqrt{\frac{4}{13}} = \frac{2}{\sqrt{13}}$

Problem 3. $\cos\theta = 2\cos^2\frac{\theta}{2} - 1$

Solution: RHS= $2\cos^2\frac{\theta}{2} - 1$

$= 2\left(\sqrt{\frac{1+\cos\theta}{2}}\right)^2 - 1$

$= 2\times\frac{1+\cos\theta}{2} - 1$

$= 1+\cos\theta - 1$

$= \cos\theta$ = LHS.

Problem 4. $\left(\sin\frac{\theta}{2} + \cos\frac{\theta}{2}\right)^2 = 1+\sin\theta$

Solution: LHS = $\left(\sin\frac{\theta}{2}+\cos\frac{\theta}{2}\right)^2$

$$=\sin^2\frac{\theta}{2}+\cos^2\frac{\theta}{2}+2\sin\frac{\theta}{2}\cos\frac{\theta}{2}$$

$$=1+2\sqrt{\frac{1+\cos\theta}{2}}.\sqrt{\frac{1+\cos\theta}{2}}$$

$$=1+\frac{2\sqrt{1-\cos^2\theta}}{\sqrt{2^2}}$$

$$=1+\frac{2\sqrt{\sin^2\theta}}{2}=1+\sin\theta = \text{RHS.}$$

Problem 5. Reduce $sin^4\theta$ to an expression involving only functions of multiples of θ, raised to the first power.

Solution: LHS = $\boldsymbol{sin^4\theta}$

$$=\left(\sin^2\theta\right)^2=\left(\frac{1-\cos 2\theta}{2}\right)^2$$

$$=\frac{1}{4}\left[1+\cos^2 2\theta-2\cos 2\theta\right]$$

$$=\frac{1}{4}\left[1+\frac{1-\cos 2\theta}{2}-2\cos 2\theta\right]$$

$$=\frac{1}{8}\left[2+1-\cos 2\theta-4\cos 2\theta\right]$$

$$=\frac{1}{8}\left[3-\cos 2\theta-4\cos 2\theta\right]$$

$$=\frac{3}{8}-\frac{1}{8}\cos 2\theta-\frac{1}{2}\cos 2\theta\cdot$$

Problem 6. Compute the value of $sin\frac{\pi}{12}$ from $\frac{\pi}{6}$

Solution: $\sin^2\frac{\pi}{12}=\frac{1-\cos\frac{\pi}{6}}{2}$

$$=\frac{1-\frac{\sqrt{3}}{2}}{2}$$

$$= \frac{2 - \sqrt{3}}{4}$$

$$\therefore \sin\frac{\pi}{12} = \frac{2 - \sqrt{3}}{4}..$$

Problem 7. $\frac{1 - \cos\alpha}{\sin\alpha} = \tan\frac{\alpha}{2}$

Solution: LHS = $\frac{1 - \cos\alpha}{2}$

$$= \frac{2\sin^2\frac{\alpha}{2}}{2\sin\frac{\alpha}{2}\cos\frac{\alpha}{2}}$$

$$= \frac{\sin\frac{\alpha}{2}}{\cos\frac{\alpha}{2}}$$

$$= \tan\frac{\alpha}{2}.$$

Problem 8. $\sqrt{\frac{1 + \sin\alpha}{1 - \sin\alpha}} = \frac{\sin\frac{\alpha}{2} + \cos\frac{\alpha}{2}}{\sin\frac{\alpha}{2} - \cos\frac{\alpha}{2}}$

Solution: LHS $\sqrt{\frac{1 + \sin\alpha}{1 - \sin\alpha}}$

$$= \sqrt{\frac{\sin^2\frac{\alpha}{2} + \cos^2\frac{\alpha}{2} + 2\sin\frac{\alpha}{2}\cos\frac{\alpha}{2}}{\sin^2\frac{\alpha}{2} + \cos^2\frac{\alpha}{2} - 2\sin\frac{\alpha}{2}\cos\frac{\alpha}{2}}}$$

$$= \sqrt{\frac{\left(\sin\frac{\alpha}{2} + \cos\frac{\alpha}{2}\right)^2}{\left(\sin\frac{\alpha}{2} - \cos\frac{\alpha}{2}\right)^2}}$$

$$= \frac{\sin\frac{\alpha}{2} + \cos\frac{\alpha}{2}}{\sin\frac{\alpha}{2} - \cos\frac{\alpha}{2}} = RHS.$$

Problem 9. Prove that $\frac{\sin A + \sin 2A}{1 + \cos A + \cos 2A} = \tan A$

Solution: LHS $= \dfrac{\sin A + \sin 2A}{1 + \cos A + \cos 2A}$

$= \dfrac{\sin A + 2\sin A \cos A}{1 + \cos A + 2\cos^2 A - 1}$

$= \dfrac{\sin A\,(1 + 2\cos A)}{\cos A\,(1 + 2\cos A)}$

$= \tan A.$

TRIPLE ANGLE IDENTITIES

$\sin 3\alpha = 3\sin\alpha - 4\sin^3\alpha$

$\cos 3\alpha = 4\cos^3\alpha - 3\cos\alpha$

$\tan 3\alpha = \dfrac{3\tan\alpha - 4\sin^3\alpha}{1 - 3\tan^2\alpha}$.

Proof:

$\sin 3\alpha = \sin(2\alpha + \alpha)$

$= \sin 2\alpha \cos\alpha + \cos 2\alpha \sin\alpha$

$= 2\sin\alpha\ \cos\alpha.\cos\alpha + \left(1 - 2\sin^2\alpha\right)\sin\alpha$

$= 2\sin\alpha\ \cos^2\alpha + \sin\alpha - 2\sin^3\alpha$

$= 2\sin\alpha\left(1 - \sin^2\alpha\right) + \sin\alpha - 2\sin^3\alpha$

$= 2\sin\alpha - 2\sin^3\alpha + \sin\alpha - 2\sin^3\alpha$

$= 3\,sin\alpha - 4\sin^3\alpha$.

$\cos 3\alpha = \cos(2\alpha + \alpha)$

$= \cos 2\alpha\ \cos\alpha - \sin 2\alpha \sin\alpha$

$= \left(2\cos^2\alpha - 1\right)\cos\alpha - (2\,sin\alpha \cos\alpha)\sin\alpha$

$= 2\cos^3\alpha - \cos\alpha - 2\sin^2\alpha\cos\alpha$

$= 2\cos^3\alpha - \cos\alpha - 2\left(1-\cos^2\alpha\right)\cos\alpha$

$= 2\cos^3\alpha - \cos\alpha - 2\cos\alpha + 2\cos^3\alpha$

$= 4\cos^3\alpha - 3\cos\alpha$.

$\tan 3\alpha = \tan(2\alpha + \alpha)$

$= \dfrac{\tan 2\alpha + \tan\alpha}{1 - \tan 2\alpha \tan\alpha}$

$= \dfrac{\left(\dfrac{\tan 2\alpha}{1-\tan^2\alpha}\right) + \tan\alpha}{1 - \left(\dfrac{2\tan}{1-\tan^2\alpha}\right)\tan\alpha}$

$= \dfrac{2\tan\alpha + \tan\alpha - \tan^3\alpha}{1 - \tan^2\alpha - 2\tan^2\alpha}$

$= \dfrac{3\tan\alpha - \tan^3\alpha}{1 - 3\tan^2\alpha}$.

Using triple angle formulas to verify identities

Problem 1. $\dfrac{\sin 3\theta}{\sin\theta} - \dfrac{\cos 3\theta}{\cos\theta} = 2$

Solution: LHS $= \dfrac{\sin 3\theta}{\sin\theta} - \dfrac{\cos 3\theta}{\cos\theta}$

$= \dfrac{\sin 3\theta\cos\theta - \cos 3\theta\sin\theta}{\sin\theta\cos\theta}$

$= \dfrac{\sin(3\theta - \theta)}{\sin\theta\cos\theta}$

$= \dfrac{\sin 2\theta}{\sin\theta\cos\theta}$

$= \dfrac{2\sin\theta\cos\theta}{\sin\theta\cos\theta}$

$= 2 =$ RHS.

Problem 2: Prove the following identity $\frac{\sin 3\theta}{\cos\theta} + \frac{\cos 3\theta}{\sin\theta} = 2\cot 2\theta$

Solution: LHS $= \frac{\sin 3\theta}{\cos\theta} + \frac{\cos 3\theta}{\sin\theta}$

$$= \frac{\sin 3\theta \sin\theta + \cos 3\theta \cos\theta}{\sin\theta\cos\theta}$$

$$= \frac{\cos(3\theta - \theta)}{\sin\theta\, cos\theta}$$

$$= \frac{\cos 2\theta}{\sin\theta\cos\theta}$$

$$= 2\frac{\cos 2\theta}{2\sin\theta\cos\theta}$$

$$= 2\frac{\cos 2\theta}{\sin 2\theta}$$

$= 2\cot 2\theta$.

At the end of this section, here is an example where all formulas; double angle, half angle and triple angle will be used along basic identities.

Problem 3: Find the values of $\sin\theta$ and $\cos\theta$ without using a table or calculator, when θ is equal to:

18°

36°

54°

72°

Also prove that $\cos 36^{\circ} \cos 72^{\circ} \cos 108^{\circ} \cos 144^{\circ} = \frac{1}{16}$.

Solution:

Let $\theta = 18^\circ$

$\Rightarrow 5\theta = 90^\circ$

$\Rightarrow 3\theta + 2\theta = 90^\circ$

$\Rightarrow 2\theta = 90^0 - 3\theta$

$\Rightarrow \sin 2\theta = \sin\left(90^\circ - 3\theta\right)$

$\Rightarrow 2\sin\theta\cos\theta = \cos 3\theta$

$\Rightarrow 2\sin\theta\cos\theta = 4\cos^3\theta - 3\cos\theta$

$\Rightarrow 2\sin\theta = 4\cos^2\theta - 3$

$\Rightarrow 2\sin\theta = 4\left(1 - \sin^2\theta\right) - 3$

$\Rightarrow 2\sin\theta = 4 - 4\sin^2\theta - 3$

$\Rightarrow 4\sin^2\theta + 2\sin\theta - 1 = 0$.

This is a quadratic equation with $a = 4$, $b = 2$ and $c = -1$.

$$\sin\theta = \frac{-2 \pm \sqrt{4+16}}{8}$$

$$\sin\theta = \frac{-2 \pm \sqrt{20}}{8}$$

$$\sin\theta = \frac{-2 \pm 2\sqrt{5}}{8}$$

$$\sin\theta = \frac{-1 \pm \sqrt{5}}{4}$$

Since $\theta = 18^\circ$ lies in first quadrant;

$$\sin\theta = \frac{-1 + \sqrt{5}}{4}$$

$$\sin 18^\circ = \frac{\sqrt{5} - 1}{4}$$

$\cos^2 18^\circ = 1 - \sin^2 18^\circ$

$$=1-\left(\frac{\sqrt{5}-1}{4}\right)^2$$

$$=1-\left(\frac{\left(\sqrt{5}\right)^2+1-2\sqrt{5}}{16}\right)$$

$$=1-\left(\frac{6-2\sqrt{5}}{16}\right)$$

$$=\frac{16-6+2\sqrt{5}}{16}$$

$$\cos 18^\circ=\sqrt{\frac{10+2\sqrt{5}}{16}}$$

$$\cos 18^\circ=\frac{\sqrt{10+2\sqrt{5}}}{4}.$$

ii.

Since $\cos^2\theta=\frac{1+\cos 2\theta}{2}$

$\cos 2\theta=2\cos^2\theta-1$.

Replace $\theta=18^\circ$ in the above equation to obtain

$$\cos 2\left(18^\circ\right)=2\cos^2 18^\circ-1$$

$$\cos 36^\circ=2\left(\frac{10+2\sqrt{5}}{16}\right)-1.$$

After simplifying we get:

$$\cos 36^\circ=\frac{1+\sqrt{5}}{2}$$

Similarly, $\sin 36^\circ=\frac{\sqrt{10-2\sqrt{5}}}{4}$.

iii. Now

$$\sin(90^\circ-36^\circ)=\cos 36^\circ$$

$$\Rightarrow \sin 54^\circ=\cos 36^\circ$$

$\Rightarrow \quad \sin 54^\circ = \frac{1+\sqrt{5}}{2}$

$\cos(90^\circ - 36^\circ) = \sin 36^\circ$

$\cos 54^\circ = \sin 36^\circ$

$\cos 54^\circ = \frac{\sqrt{10-2\sqrt{5}}}{4}.$

iv. For 72°

$\sin(90^\circ - 18^\circ) = \cos 18^\circ$

$\Rightarrow \sin 72^\circ = \cos 18^\circ$

$\Rightarrow \quad \sin 72^\circ = \frac{\sqrt{10+2\sqrt{5}}}{4}$

$\cos\left(90^\circ - 18^\circ\right) = \sin 18^\circ$

$\Rightarrow \quad \cos 72^\circ = \frac{\left(\sqrt{5}-1\right)}{4}$

$\cos 108^\circ = \cos(90^\circ + 18^\circ)$

$\cos 108^\circ = -\sin 18^\circ = \frac{1-\sqrt{5}}{4}$

$\cos 144^\circ = \cos(180^\circ - 36^\circ) = -\cos 36^\circ = \frac{-1-\sqrt{5}}{4}.$

Now use all above results in the equation i.e.

$= \cos 36^\circ \cos 72^\circ \cos 108^\circ \cos 144^\circ$

$= \left(\frac{1+\sqrt{5}}{2}\right)\left(\frac{\left(\sqrt{5}-1\right)}{4}\right)\left(\frac{1-\sqrt{5}}{4}\right)\left(\frac{-1-\sqrt{5}}{4}\right)$

$= \frac{1}{16} = \text{RHS.}$

SUM, DIFFERENCES AND PRODUCT OF SINE AND COSINES

We have already learned a number of formulas useful for expanding or simplifying trigonometric expressions, but sometimes we may need to express the product of the cosine and sine as a sum. We can use the product-to-sum formulas, which express products of trigonometric functions as sums.

We have previously discussed some formulas

$$\sin(\alpha+\beta) = \sin\alpha\cos\beta + \cos\alpha\sin\beta \tag{1}$$

$$\sin(\alpha-\beta) = \sin\alpha\cos\beta - \cos\alpha\sin\beta \tag{2}$$

$$\cos(\alpha+\beta) = \cos\alpha\cos\beta - \sin\alpha\sin\beta \tag{3}$$

$$\cos(\alpha-\beta) = \cos\alpha\cos\beta + \sin\alpha\sin\beta \tag{4}$$

Adding (1) and (2) we get

$$\sin(\alpha+\beta) + \sin(\alpha-\beta) = 2\sin\alpha\cos\beta \tag{5}$$

Subtracting (2) from (1) we get

$$\sin(\alpha+\beta) - \sin(\alpha-\beta) = 2\cos\alpha\sin\beta \tag{6}$$

Adding (3) and (4) we get

$$\cos(\alpha+\beta) + \cos(\alpha-\beta) = 2\cos\alpha\cos\beta \tag{7}$$

Subtracting (4) from (3) we get

$$\cos(\alpha+\beta) - \cos(\alpha-\beta) = -2\sin\alpha\sin\beta \tag{8}$$

Thus, we have four identities, as mentioned above in results (5), (6), (7) and (8).

Conversion from product to sum/difference

$$\sin(\alpha+\beta) + \sin(\alpha-\beta) = 2\sin\alpha\cos\beta$$

$$\sin(\alpha+\beta) - \sin(\alpha-\beta) = 2\cos\alpha\sin\beta$$

$\cos(\alpha+\beta)+\cos(\alpha-\beta)=2\cos\alpha\cos\beta$

$\cos(\alpha+\beta)-\cos(\alpha-\beta)=-2\sin\alpha\sin\beta$

Now, replace $\alpha+\beta=A$ and $\alpha-\beta=B$, we obtain

$\alpha=\frac{A+B}{2}$ and $\beta=\frac{A-B}{2}$.

Thus, we can write formulas

Conversion from sum/difference to product

$$\sin A+\sin B=2\sin\frac{A+B}{2}\cos\frac{A-B}{2}$$

$$\sin A-\sin B=2\cos\frac{A+B}{2}\sin\frac{A-B}{2}$$

$$\cos A+\cos B=2\cos\frac{A+B}{2}\cos\frac{A-B}{2}$$

$$\cos A-\cos B=-2\sin\frac{A+B}{2}\sin\frac{A-B}{2}$$

Problem 1. Express the following products as sums or differences

$\cos(2x+30^\circ)\cos(2x-30^\circ)$

$\sin 12^0 \sin 46^\circ$

$\sin(x+45^\circ)\sin(x-45^\circ)$

$\cos(x+y)\sin(x-y)\cdot$

Solution:

$\cos(2x+30^\circ)\cos(2x-30^\circ)$

$=\frac{1}{2}\left(2\cos(2x+30^\circ)\cos(2x-30^\circ)\right)$

$=\frac{1}{2}\left(\cos(2x+30^\circ+2x-30^\circ)+\cos(2x+30^\circ-2x+30^\circ)\right)$

$$=\frac{1}{2}\left(\cos 4x+\cos 60^{\circ}\right).$$

$\sin 12^{\circ} \sin 46^{\circ}$

$$=-\frac{1}{2}(-2 \sin 12^{\circ} \sin 46^{\circ}$$

$$=-\frac{1}{2}\left(\cos(12^{\circ}+46^{\circ}\right)-\cos(12^{\circ}-46^{\circ}))$$

$$=-\frac{1}{2}\left(\cos 58^{\circ}-\cos\left(-34^{\circ}\right)\right)$$

$$=-\frac{1}{2}\left(\cos 58^{\circ}-\cos 34^{\circ}\right)\cdot$$

$\sin(x+45^{\circ})\sin(x-45^{\circ})$

$$=-\frac{1}{2}\left(-2\sin(x+45^{\circ}\right)\sin(x-45^{\circ}))$$

$$=-\frac{1}{2}\left(\cos(x+45^{\circ}+x-45^{\circ}\right)-\cos(x+45^{\circ}-x+45^{\circ})$$

$$=-\frac{1}{2}\left(\cos(2x\right)-\cos(90^{\circ}).$$

$\cos(x+y)\sin(x-y)$

$$=\frac{1}{2}\left(\sin(x+y+x-y\right)-\sin(x+y-x+y))$$

$$=\frac{1}{2}\left(\sin(2x\right)-\sin(2y)).$$

Problem 2. Express the following sums or differences as product.

$\sin 5\theta+\sin 3\theta$

$\cos 12^{\circ}+\cos 48^{\circ}$

$\cos 7\theta - \cos\theta$

$\sin(x+30^\circ)+\sin(x-30^\circ)$.

Solution:

$\sin 5\theta + \sin 3\theta$

$$= 2\sin(\frac{5\theta+3\theta}{2})\cos(\frac{5\theta-3\theta}{2})$$
$$= 2\sin 4\theta\cos\theta.$$

$\cos 12^\circ + \cos 48^\circ$

$$= 2\cos\left(\frac{12^\circ+48^\circ}{2}\right)\cos(\frac{12^\circ-48^\circ}{2})$$
$$= 2\cos 30^\circ\cos(-18^\circ)$$
$$= 2\cos 30^\circ\cos 18^\circ.$$

$\cos 7\theta - \cos\theta$

$$= -2\sin(\frac{7\theta+\theta}{2})\sin(\frac{7\theta-\theta}{2})$$
$$= -2\sin 4\theta\sin 3\theta.$$

$\sin(x+30^\circ)+\sin(x-30^\circ)$

$$= 2\sin(\frac{x+30^\circ+x-30^\circ}{2})\cos(\frac{x+30^\circ-x+30^\circ}{2})$$
$$= 2\sin(x)\cos(30^\circ).$$

Problem 3. Express $\sin 3\theta + \sin 5\theta + \sin 7\theta + \sin 9\theta$ **as a product.**

solution : $\sin 3\theta + \sin 5\theta + \sin 7\theta + \sin 9\theta$

$$= (\sin 9\theta + \sin 3\theta) + (\sin 7\theta + \sin 5\theta)$$

$$= 2\sin\left(\frac{9\theta + 3\theta}{2}\right)\cos\left(\frac{9\theta - 3\theta}{2}\right) + 2\sin\left(\frac{7\theta + 5\theta}{2}\right)\cos\left(\frac{7\theta - 5\theta}{2}\right)$$

$$= 2\sin\left(\frac{12\theta}{2}\right)\cos\left(\frac{6\theta}{2}\right) + 2\sin\left(\frac{12\theta}{2}\right)\cos\left(\frac{2\theta}{2}\right)$$

$$= 2\sin 6\theta \cos 3\theta + 2\sin 6\theta \cos\theta$$

$$= 2\sin 6\theta(\cos 3\theta + \cos\theta)$$

$$= 2\sin 6\theta\left(2\cos\left(\frac{3\theta + \theta}{2}\right)\cos\left(\frac{3\theta - \theta}{2}\right)\right)$$

$$= 2\sin 6\theta\left(2\cos\left(\frac{4\theta}{2}\right)\cos\left(\frac{2\theta}{2}\right)\right)$$

$$= 2\sin 6\theta(2\cos 2\theta \cos\theta)$$

$$= 4\sin 6\theta \cos 2\theta \cos\theta.$$

Problem 4. $\dfrac{\sin 5\theta - \sin 3\theta}{\sin 5\theta + \sin 3\theta} = \tan\theta$

solution : LHS $= \dfrac{\sin 5\theta - \sin 3\theta}{\sin 5\theta + \sin 3\theta}$

$$= \frac{2\cos\left(\frac{5\theta + 3\theta}{2}\right)\sin\left(\frac{5\theta - 3\theta}{2}\right)}{2\cos\left(\frac{5\theta + 3\theta}{2}\right)\cos\left(\frac{5\theta - 3\theta}{2}\right)}$$

$$= \frac{\cos\left(\frac{6\theta}{2}\right)\sin\left(\frac{2\theta}{2}\right)}{\cos\left(\frac{6\theta}{2}\right)\cos\left(\frac{2\theta}{2}\right)} = \frac{\sin\theta}{\cos\theta}$$

$= \tan\theta =$ RHS.

Problem 5. $\dfrac{\sin 5\theta + \sin 3\theta}{\sin 5\theta - \sin 3\theta} = -\cot\theta$

solution : LHS= $\dfrac{\sin 5\theta + \sin 3\theta}{\sin 5\theta - \sin 3\theta}$

$= \dfrac{2\sin\left(\dfrac{5\theta+3\theta}{2}\right)\cos\left(\dfrac{5\theta-3\theta}{2}\right)}{-2\sin\left(\dfrac{5\theta+3\theta}{2}\right)\sin\left(\dfrac{5\theta-3\theta}{2}\right)}$

$= -\dfrac{\cos\left(\dfrac{2\theta}{2}\right)}{\sin\left(\dfrac{2\theta}{2}\right)}$

$= -\dfrac{\cos\theta}{\sin\theta}$

$= -\cot\theta$ = RHS.

Problem 6. $\dfrac{\cos\beta + \cos 9\beta}{\sin\beta + \sin 9\beta} = \cot 5\beta$

solution : LHS = $\dfrac{\cos\beta + \cos 9\beta}{\sin\beta + \sin 9\beta}$

$= \dfrac{2\cos\left(\dfrac{\beta+9\beta}{2}\right)\cos\left(\dfrac{\beta-9\beta}{2}\right)}{2\sin\left(\dfrac{\beta+9\beta}{2}\right)\cos\left(\dfrac{\beta-9\beta}{2}\right)}$

$= \dfrac{\cos\left(\dfrac{10\beta}{2}\right)}{\sin\left(\dfrac{10\beta}{2}\right)}$

$= \dfrac{\cos 5\beta}{\sin 5\beta}$

$= \cot 5\beta$ = RHS.

Problem 7. $\dfrac{\sin 3\theta - \sin\theta}{\cos^2\theta + \sin^2\theta} = 2\sin\theta$

solution : LHS= $\dfrac{\sin 3\theta - \sin\theta}{\cos^2\theta - \sin^2\theta}$

$= \dfrac{2\cos\left(\dfrac{3\theta+\theta}{2}\right)\sin\left(\dfrac{3\theta-\theta}{2}\right)}{\cos 2\theta}$

$$= \frac{2\cos\left(\frac{4\theta}{2}\right)\sin\left(\frac{2\theta}{2}\right)}{\cos 2\theta}$$

$$= \frac{2\cos 2\theta \sin\theta}{\cos 2\theta}$$

$= 2\sin\theta =$ RHS.

Problem 8. $\frac{\sin \mathbf{A} + \sin \mathbf{B}}{\cos \mathbf{A} + \cos \mathbf{B}} = \tan\frac{\mathbf{A}+\mathbf{B}}{2}$

solution : LHS $= \frac{\sin A + \sin B}{\cos A + \cos B}$

$$= \frac{2\sin\left(\frac{A+B}{2}\right)\cos\left(\frac{A-B}{2}\right)}{2\cos\left(\frac{A+B}{2}\right)\cos\left(\frac{A-B}{2}\right)}$$

$$= \frac{\sin\left(\frac{A+B}{2}\right)}{\cos\left(\frac{A+B}{2}\right)} = \tan\frac{A+B}{2}$$

= RHS.

Problem 9. $\frac{\cos 2\theta - \cos 6\theta}{\cos 2\theta + \cos 6\theta} = \tan 4\theta \tan 2\theta$

solution : LHS $= \frac{\cos 2\theta - \cos 6\theta}{\cos 2\theta + \cos 6\theta}$

$$= \frac{-2\sin\left(\frac{2\theta+6\theta}{2}\right)\sin\left(\frac{2\theta-6\theta}{2}\right)}{2\cos\left(\frac{2\theta+6\theta}{2}\right)\cos\left(\frac{2\theta-6\theta}{2}\right)}$$

$$= -\frac{\sin\left(\frac{8\theta}{2}\right)\sin\left(-\frac{4\theta}{2}\right)}{\cos\left(\frac{8\theta}{2}\right)\cos\left(\frac{4\theta}{2}\right)}$$

$$= \frac{\sin 4\theta \sin 2\theta}{\cos 4\theta \cos 2\theta}$$

$= \tan 4\theta \tan 2\theta =$ RHS.

Problem 10. $\dfrac{\cos\alpha-\cos\beta}{\cos\alpha+\cos\beta}=-\dfrac{\tan\left(\dfrac{\alpha+\beta}{2}\right)}{\cot\left(\dfrac{\alpha-\beta}{2}\right)}$

solution : LHS $=\dfrac{\cos\alpha-\cos\beta}{\cos\alpha+\cos\beta}$

$$=\frac{-2\sin\left(\frac{\alpha+\beta}{2}\right)\sin\left(\frac{\alpha-\beta}{2}\right)}{2\cos\left(\frac{\alpha+\beta}{2}\right)\cos\left(\frac{\alpha-\beta}{2}\right)}$$

$$=-\tan\left(\frac{\alpha+\beta}{2}\right)\tan\left(\frac{\alpha-\beta}{2}\right)$$

$$=-\tan\left(\frac{\alpha+\beta}{2}\right)\frac{1}{\cot\left(\frac{\alpha-\beta}{2}\right)}$$

$$-\frac{\tan\left(\frac{\alpha+\beta}{2}\right)}{\cot\left(\frac{\alpha-\beta}{2}\right)}=\text{RHS.}$$

Problem 11. $\dfrac{\sin\theta+\sin 2\theta+\sin 3\theta}{\cos\theta+\cos 2\theta+\cos 3\theta}=\tan 2\theta$

solution : LHS $=\dfrac{\sin\theta+\sin 2\theta+\sin 3\theta}{\cos\theta+\cos 2\theta+\cos 3\theta}$

$$=\frac{\sin 3\theta+\sin\theta+\sin 2\theta}{\cos 3\theta+\cos\theta+\cos 2\theta}$$

$$=\frac{2\sin\left(\frac{3\theta+\theta}{2}\right)\cos\left(\frac{3\theta-\theta}{2}\right)+\sin 2\theta}{2\cos\left(\frac{3\theta+\theta}{2}\right)\cos\left(\frac{3\theta-\theta}{2}\right)+\cos 2\theta}$$

$$=\frac{\sin\left(\frac{4\theta}{2}\right)+\sin 2\theta}{\cos\left(\frac{4\theta}{2}\right)+\cos 2\theta}$$

$$=\frac{\sin(2\theta)+\sin 2\theta}{\cos(2\theta)+\cos 2\theta}$$

$$= \frac{2\sin\left(\frac{2\theta+2\theta}{2}\right)\cos\left(\frac{2\theta-2\theta}{2}\right)}{2\cos\left(\frac{2\theta+2\theta}{2}\right)\cos\left(\frac{2\theta-2\theta}{2}\right)}$$

$$= \frac{\sin 2\theta}{\cos 2\theta} = \tan 2\theta = \text{RHS.}$$

Problem 12. $\sin 5\theta + 2\sin 3\theta + \sin\theta = 4\sin 3\theta\cos^2\theta$

solution : LHS = $\sin 5\theta + 2\sin 3\theta + \sin\theta$

$= \sin 5\theta + \sin\theta + 2\sin 3\theta$

$= 2\sin\left(\frac{5\theta+\theta}{2}\right)\cos\left(\frac{5\theta-\theta}{2}\right) + 2\sin 3\theta$

$= 2\sin\left(\frac{6\theta}{2}\right)\cos\left(\frac{4\theta}{2}\right) + 2\sin 3\theta$

$= 2\sin 3\theta\cos 2\theta + 2\sin 3\theta$

$= 2\sin 3\theta(\cos 2\theta + 1)$

$= 2\sin 3\theta(\cos^2\theta - \sin^2\theta + 1)$

$= 2\sin 3\theta(\cos^2\theta + \cos^2\theta) = 4\sin 3\theta\cos^2\theta =$ RHS.

Problem 13. Show that:

$$\frac{\sin 75° - \sin 15°}{\cos 75° + \cos 15°} = \frac{1}{\sqrt{3}}$$

solution : LHS = $\dfrac{\sin 75° - \sin 15°}{\cos 75° + \cos 15°}$

$$= \frac{2\cos\left(\frac{75°+15°}{2}\right)\sin\left(\frac{75°-15°}{2}\right)}{2\cos\left(\frac{75°+15°}{2}\right)\cos\left(\frac{75°-15°}{2}\right)}$$

$$= \frac{\sin\left(\frac{60^\circ}{2}\right)}{\cos\left(\frac{60^\circ}{2}\right)}$$

$$= \frac{\sin 30^\circ}{\cos 30^\circ}$$

$$= \frac{\frac{1}{2}}{\frac{\sqrt{3}}{2}} = \frac{1}{\sqrt{3}} = \text{RHS.}$$

$\sin 20^\circ + \sin 40^\circ = \cos 15^\circ$

solution : LHS = $\sin 20^\circ + \sin 40^\circ$

$$= 2\sin\left(\frac{20^\circ + 40^\circ}{2}\right)\cos\left(\frac{20^\circ - 40^\circ}{2}\right)$$

$$= 2\sin\left(\frac{60^\circ}{2}\right)\cos\left(-\frac{20^\circ}{2}\right)$$

$$= 2\sin 30^\circ \cos 10^\circ$$

$$= 2\left(\frac{1}{2}\right)\cos 10^\circ = \cos 10^\circ = \text{RHS.}$$

$\cos 80^\circ + \cos 40^\circ = \cos 20^\circ$

solution : LHS = $\cos 80^\circ + \cos 40^\circ$

$$= 2\cos\left(\frac{80^\circ + 40^\circ}{2}\right)\cos\left(\frac{80^\circ - 40^\circ}{2}\right)$$

$$= 2\cos\left(\frac{120^\circ}{2}\right)\cos\left(\frac{40^\circ}{2}\right)$$

$$= 2\sin 60^\circ \cos 20^\circ$$

$$= 2\left(\frac{1}{2}\right)\cos 10^\circ = \cos 20^\circ = \text{RHS.}$$

$\cos 20^\circ + \cos 100^\circ + \cos 140^\circ = 0$

solution : LHS = $\cos 20^\circ + \cos 100^\circ + \cos 140^\circ$

$$= \cos 100^\circ + \cos 140^\circ + \cos 20^\circ$$

$$= 2\cos\left(\frac{100^\circ + 140^\circ}{2}\right)\cos\left(\frac{100^\circ - 140^\circ}{2}\right) + \cos 20^\circ$$

$$= 2\cos\left(\frac{240°}{2}\right)\cos\left(\frac{-40°}{2}\right) + \cos 20°$$

$$= 2\cos 120° \cos 20° + \cos 20°$$

$$= 2\left(-\frac{1}{2}\right)\cos 20° + \cos 20°$$

$$= -\cos 20° + \cos 20° = 0 = \text{RHS.}$$

Problem 14. Prove that: $\sin 20° \sin 40° \sin 60° \sin 80° = \frac{3}{16}$

Solution : LHS= $\sin 20° \sin 40° \sin 60° \sin 80°$

$$= \frac{\sqrt{3}}{2}[\sin 80° \sin 40° \sin 20°]$$

$$= \frac{\sqrt{3}}{2}\left[\frac{-2}{-2}\sin 80° \sin 40°\right]\sin 20°$$

$$= -\frac{\sqrt{3}}{4}[-2\sin 80° \sin 40°]\sin 20°$$

$$= -\frac{\sqrt{3}}{4}[\cos(80° + 40°) - \cos(80° - 40°)]\sin 20°$$

$$= -\frac{\sqrt{3}}{4}[\cos 120° - \cos 40°]\sin 20°$$

$$= -\frac{\sqrt{3}}{4}\left[\frac{-1}{2} - \cos 40°\right]\sin 20°$$

$$= -\frac{\sqrt{3}}{8}\sin 20° + \frac{\sqrt{3}}{4}\cos 40° \sin 20°$$

$$= \frac{\sqrt{3}}{8}\sin 20° + \frac{\sqrt{3}}{8}(2\cos 40° \sin 20°)$$

$$= \frac{\sqrt{3}}{8}\sin 20° + \frac{\sqrt{3}}{8}(\sin(40° + 20°) - \sin(40° - 20°))$$

$$=\frac{\sqrt{3}}{8}\sin 20^\circ+\frac{\sqrt{3}}{8}[\sin 60^\circ-\sin 20^\circ]$$

$$=\frac{\sqrt{3}}{8}\sin 20^\circ+\frac{\sqrt{3}}{8}\left[\frac{\sqrt{3}}{2}-\sin 20^\circ\right]$$

$$=\frac{\sqrt{3}}{8}\sin 20^\circ+\frac{3}{16}-\frac{\sqrt{3}}{8}\sin 20^\circ=\frac{3}{16}=\text{RHS.}$$

Problem 15. $\sin 10^\circ \sin 30^\circ \sin 50^\circ \sin 70^\circ=\frac{1}{16}$

Solution: LHS = $\sin 10^\circ \sin 30^\circ \sin 50^\circ \sin 70^\circ$

$$=\frac{1}{2}[\sin 70^\circ \sin 50^\circ \sin 10^\circ]$$

$$=\frac{1}{2}\left[\frac{-2}{-2}\sin 70^\circ \sin 50^\circ\right]\sin 10^\circ$$

$$=-\frac{1}{4}[-2\sin 70^\circ \sin 50^\circ]\sin 10^\circ$$

$$=-\frac{1}{4}\left[\cos\left(70^\circ+50^\circ\right)-\cos\left(70^\circ-50^\circ\right)\right]\sin 10^\circ$$

$$=-\frac{1}{4}[\cos 120^\circ-\cos 20^\circ]\sin 10^\circ$$

$$=-\frac{1}{4}\left[\frac{-1}{2}-\cos 20^\circ\right]\sin 10^\circ$$

$$=\frac{1}{8}\sin 10^\circ+\frac{\sqrt{3}}{4}\cos 20^\circ \sin 10^\circ$$

$$=\frac{1}{8}\sin 10^\circ+\frac{1}{8}(2\cos 20^\circ \sin 10^\circ)$$

$$=\frac{1}{8}\sin 10^\circ+\frac{1}{8}(\sin(20^\circ+10^\circ)-\sin(20^\circ-10^\circ))$$

$$=\frac{1}{8}\sin 10^\circ+\frac{1}{8}[\sin 30^\circ-\sin 10^\circ]$$

$$=\frac{1}{8}\sin 10^\circ+\frac{1}{8}\left[\frac{1}{2}-\sin 10^\circ\right]$$

$$=\frac{1}{8}\sin 10^\circ+\frac{1}{16}-\frac{1}{8}\sin 10^\circ=\frac{1}{16}= \text{RHS.}$$

Problem 16. $\sin 20^\circ \sin 40^\circ \sin 80^\circ \sin 90^\circ=\frac{\sqrt{3}}{8}$

Solution: LHS= $\sin 20^\circ \sin 40^\circ \sin 80^\circ \sin 90^\circ$

$$=1[\sin 80^\circ \sin 40^\circ \sin 20^\circ]$$

$$=1\left[\frac{-2}{-2}\sin 80^\circ \sin 40^\circ\right]\sin 20^\circ$$

$$=-\frac{1}{2}[-2\sin 80^\circ \sin 40^\circ]\sin 20^\circ$$

$$=-\frac{1}{2}[\cos(80^\circ+40^\circ)-\cos(80^\circ-40^\circ)]\sin 20^\circ$$

$$=-\frac{1}{2}[\cos 120^\circ-\cos 40^\circ]\sin 20^\circ$$

$$=-\frac{1}{2}\left[\frac{-1}{2}-\cos 40^\circ\right]\sin 20^\circ$$

$$=\frac{1}{4}\sin 20° + \frac{1}{2}\cos 40° \sin 20°$$

$$=\frac{1}{4}\sin 20° + \frac{1}{4}(2\cos 40° \sin 20°)$$

$$=\frac{1}{4}\sin 20° + \frac{1}{4}(\sin(40° + 20°) - \sin(40° - 20°))$$

$$=\frac{1}{4}\sin 20° + \frac{1}{4}[\sin 60° - \sin 20°]$$

$$=\frac{1}{4}\sin 20° + \frac{1}{4}\left[\frac{\sqrt{3}}{2} - \sin 20°\right]$$

$$=\frac{1}{4}\sin 20° + \frac{\sqrt{3}}{8} - \frac{1}{4}\sin 20° = \frac{\sqrt{3}}{8} = \text{RHS.}$$

Problem 17. $\cos 20° \cos 40° \cos 60° \cos 80° = \dfrac{1}{16}$

Solution : LHS = $\cos 20° \cos 40° \cos 60° \cos 80°$

$$=\frac{1}{2}[\cos 80° \cos 40° \cos 20°]$$

$$=\frac{1}{2}\left[\frac{2}{2}\cos 80° \cos 40°\right]\cos 20°$$

$$=\frac{1}{4}[2\cos 80° \cos 40°]\cos 20°$$

$$=\frac{1}{4}\left[\cos(80° + 40°) + \cos(80° - 40°)\right]\cos 20°$$

$$=\frac{1}{4}[\cos 120° + \cos 40°]\sin 20°$$

$$=\frac{1}{4}\left[\frac{-1}{2}+\cos 40^\circ\right]\cos 20^\circ$$

$$=-\frac{1}{8}\cos 20^\circ+\frac{1}{4}\cos 40^\circ \cos 20^\circ$$

$$=-\frac{1}{8}\cos 20^\circ+\frac{1}{8}(2\cos 40^\circ \cos 20^\circ)$$

$$=-\frac{1}{8}\cos 20^\circ+\frac{1}{8}(\cos(40^\circ+20^\circ)+\cos(40^\circ-20^\circ))$$

$$=-\frac{1}{8}\cos 20^\circ+\frac{1}{8}[\cos 60^\circ-\cos 20^\circ]$$

$$=\frac{-1}{8}\cos 20^\circ+\frac{1}{8}\left[\frac{1}{2}-\cos 20^\circ\right]$$

$$=-\frac{1}{8}\cos 20^\circ+\frac{3}{16}+\frac{1}{8}\cos 20^\circ=\frac{1}{16}=\text{RHS.}$$

SUMMARY

Most important chapter for trigonometry. We learn how we can use the double angle, half angle and triple angle identities to solve different trigonometric identities. The double-angle formulas can be used to derive the reduction formulas, which we can use to reduce the power of a given expression involving even powers of sine or cosine. They allow us to rewrite the even powers of sine or cosine in terms of the first power of cosine. These formulas are especially important in higher-level math courses, calculus in particular. Also called the power-reducing formulas, these three identities are easily derived from the double-angle formulas.

The next set of identities is the set of **half-angle formulas**, which can be derived from the reduction formulas. These can be used when we have an angle that is half the size of a special angle.

The third section of the chapter is related to triple angle identities and the final section belongs to the conversion of the sum/difference to products and vice versa.

Key terms and important formulas

Double angle identities

$\sin 2\alpha = 2\sin\alpha\cos\alpha$

$\cos 2\alpha = \cos^2\alpha - \sin^2\alpha$

$\cos 2\alpha = 2\cos^2\alpha - 1$

$\cos 2\alpha = 1 - 2\sin^2\alpha$

$$\tan(2\alpha) = \frac{2\tan\alpha}{1-\tan^2\alpha}$$

Reduction formulas

$$\sin^2\theta = \frac{1-\cos 2\theta}{2}$$

$$\cos^2\theta = \frac{1+\cos 2\theta}{2}$$

$$\tan^2\theta = \frac{1-\cos 2\theta}{1+\cos 2\theta}$$

Half Angle identities

$$\sin\theta = \pm\sqrt{\frac{1-\cos 2\theta}{2}}$$

$$\cos\theta = \pm\sqrt{\frac{1+\cos 2\theta}{2}}$$

$$\tan\theta = \pm\sqrt{\frac{1-\cos 2\theta}{1+\cos 2\theta}}$$

Conversion from product to sum/difference

$\sin(\alpha+\beta) + \sin(\alpha-\beta) = 2\sin\alpha\cos\beta$

$\sin(\alpha+\beta) - \sin(\alpha-\beta) = 2\cos\alpha\sin\beta$

$$\cos(\alpha+\beta)+\cos(\alpha-\beta)=2\cos\alpha\cos\beta$$

$$\cos(\alpha+\beta)-\cos(\alpha-\beta)=-2\sin\alpha\sin\beta$$

Conversion from sum/difference to product

$$\sin A+\sin B=2\sin\frac{A+B}{2}\cos\frac{A-B}{2}$$

$$\sin A-\sin B=2\cos\frac{A+B}{2}\sin\frac{A-B}{2}$$

$$\cos A+\cos B=2\cos\frac{A+B}{2}\cos\frac{A-B}{2}$$

$$\cos A-\cos B=-2\sin\frac{A+B}{2}\sin\frac{A-B}{2}.$$

CHAPTER 5

APPLICATION OF RIGHT ANGLE TRIANGLE

INTRODUCTION

A right triangle is a triangle in which one angle is a right angle (that is, a 90-degree angle). The relationship between the sides and angles of a right triangle is the basis for trigonometry. The side opposite the right angle is called the hypotenuse. The sides adjacent to the right angle are called legs. A side a may be identified as the side adjacent to angle B and opposite to angle A, while side b is the side adjacent to angle A and opposite to angle B.

The solution of right triangles depends on the approximate values for trigonometric functions of acute angles. An important part of the solution is determining the appropriate value to use for a trigonometric function. This part of the solution is different when you are using tables a scientific calculator.

In general, the procedure involves using the given data to write an equation using a trigonometric function and then solving for the unknown value in the equation. The given data will consist of either two sides of a right triangle or of one side and an acute angle. Once one value has been found, a second acute angle and the remaining side can be found. The second acute angle is found using the fact that the acute angles of a right triangle are complementary. The third side is found using a definition of a second trigonometric function or by Pythagorean Theorem.

If the lengths of all three sides of a right triangle are integers, the triangle is said to be a Pythagorean triangle and its side lengths are collectively known as a Pythagorean triple.

ANGLES OF ELEVATION AND DEPRESSION

If $\overrightarrow{OA}$ is the horizontal ray through the eye of the observer at point O, and there are two objects B and ,C such that B is above and C is below the horizontal ray, $\overrightarrow{OA}$, then:

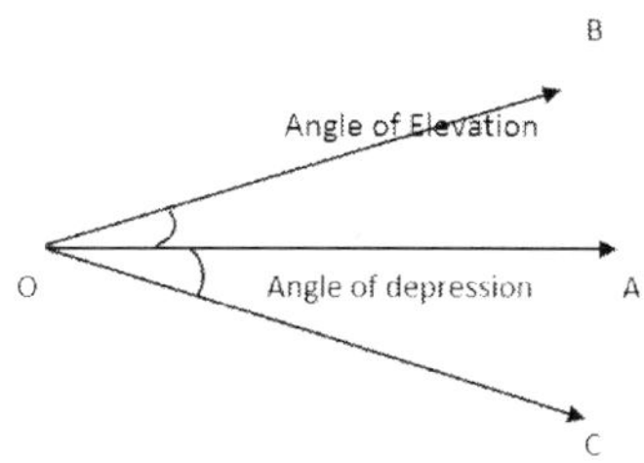

i) To look at B above the horizontal ray, we have to raise our eye, and ∠AOB is called the **Angle of Elevation.**

ii) To look at C below the horizontal ray we have to lower our eye, and ∠AOC is called the Angle **of Depression.**

Calculator values of trigonometric functions

Calculators give values of trigonometric functions based on the number of digits that can be displayed, usually 8, 10, or 12. The number of decimal places shown varies with the size of the number but is usually at least four. When a calculator is used in this book, all trigonometric function values shown will be rounded to six digits unless the value is exact using fewer digits.

Example: The string of a flying Kite is 200 meters long, and its angle of elevation is 60. Find the height of the kite above the ground, taking the string to be fully stretched.

Solution: Let be the position of the observer, be the position of the kite and be the horizontal ray through.

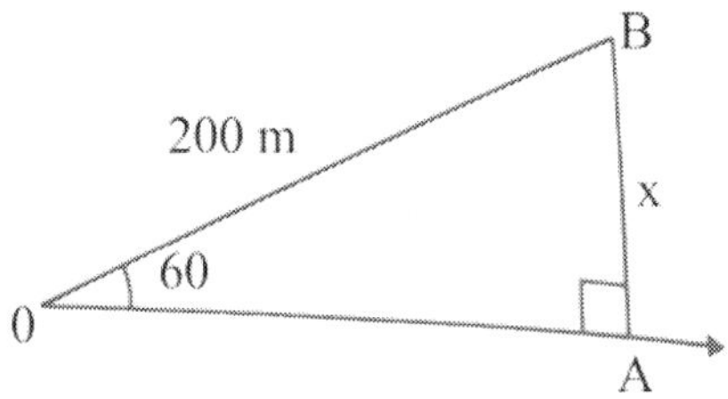

Figure 7

Draw $\overline{BA} \perp \overline{OA}$.

Now $m\angle O = 60^\circ$ and $OB = 200\,m$

Suppose $AB = x$ meters

In right $\Delta\, AOB$,

$$\frac{x}{200} = \sin 60^\circ = \frac{\sqrt{3}}{2}$$

$$\Rightarrow x = 200\left(\frac{1.732}{2}\right)$$

$x = 100(1.732) = 173.2$.

Hence, the height of the Kite above the ground is $173.2\,m$.

Example: A surveyor stand on the top of a 240 m high hill, and observes two boats at the angles of depression of measures 17° and 10° in the same straight line with the foot of the hill, just below the observers. Find the distance between the two boats, if they are on the same side of the hill.

Solution: Let T be the top of the hill $\overline{TM}$, where the observer is stationed, and A and B be positions of the two boats so that m$\angle XTB$ = 10° and m$\angle XTA$ = 17° and TM = 240m.

Now, m$\angle MAT =$ m$\angle XTA = 17^\circ$ and m $\angle MBT$ = m$\angle XTB = 10^\circ$.

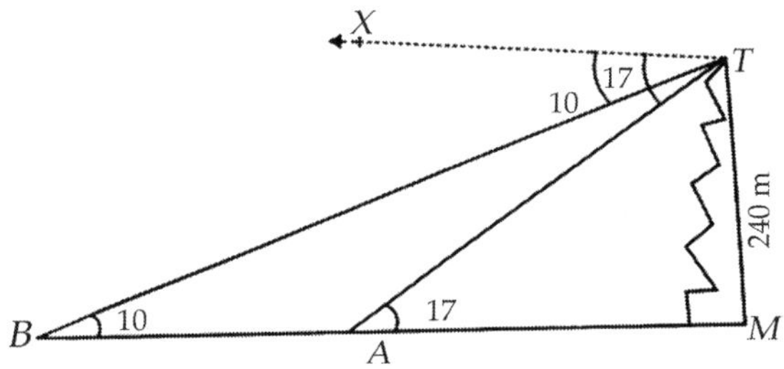

Figure 8

From the figure, $\frac{TM}{AM} = tan17°$

$\Rightarrow$ AM = $\frac{TM}{tan17°} = \frac{240}{0.3057}$

$\Rightarrow$ AM = 785 m

and $\frac{TM}{BM} = 10°$

$\Rightarrow$ BM = $\frac{TM}{tan10°}$

$BM = \frac{240}{0.1763}$ =1361 m

$\therefore AB = BM - AM = 1361 - 785 = 576m$.

Hence the distance between the boats is 576 m.

Example: From a point 100 m above the surface of a lake, the angle of elevation of a peak of a cliff is found to be 15° and the angle of depression of the image of the peak is 30°. Find the height of the peak.

Solution: Let A be the top of the peak and $\overline{AM}$ and $\overline{MB}$ be its image . Let P be the point just below P (on the surface of the lake).

For P, draw PQ $\perp$ AM

Q = h – QM = h – PL = h - 100

From the figure, we have

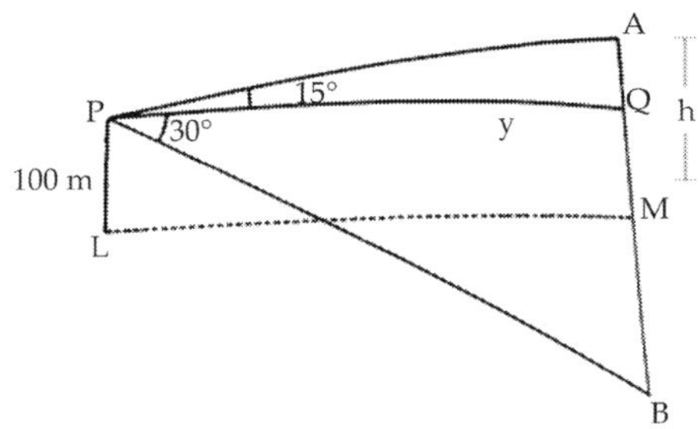

Figure 9

$$Tan\ 15° = \frac{AQ}{PQ} = \frac{h-100}{y} \text{ and } \tan 30° = \frac{BQ}{PQ} = \frac{100+h}{y}$$

By division, we get

$$\frac{\tan 15°}{\tan 30°} = \frac{h-100}{h+100}$$

By Componendo and Dividendo, we have

$$\frac{\tan 15° + \tan 30°}{\tan 15° - \tan 30°} = \frac{h-100+h+100}{h-100-h-100} = \frac{2h}{-200} = \frac{h}{-100}$$

$$h = \frac{\tan 30° + \tan 15°}{\tan 30° - \tan 15°} \times 100 = \frac{0.5774 + 0.2679}{0.5774 - 0.2679} \times 100$$

h = 273.1179.

Hence the height of the peak ≈ 273 m.

Example: An O.P. sitting on a cliff 1900 meters high finds himself in the same vertical plane with an anti-air-craft gun of the ammunition depot of the enemy, and observes that the angle of depression of the gun and the depot are 60° and 30°respectively. He passes this information on to the headquarters. Calculate the distance between the gun and the depot.

Solution: Let O be the position of the O.P., A be the point on the ground just below him and B and C be the positions of the gun and the depot respectively

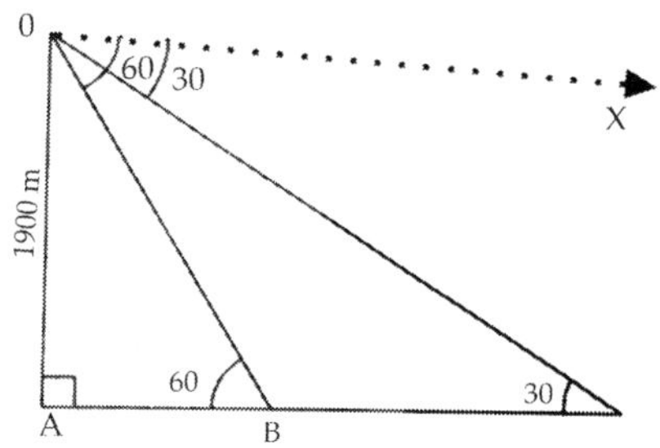

Figure 10

OA =1900m

m $\angle BOX$ = 60 °

and m $\angle COX$ = 30 °

$\Rightarrow m\angle ABO = m\angle BOX = 60^\circ$, $m\angle ACO = m\angle COX = 30^\circ$.

In right ΔBAQ.

$\frac{1900}{AB} = \tan 60^\circ$

$\Rightarrow AB = \frac{1900}{\tan 60^\circ} = \frac{1900}{\sqrt{3}}$

$\frac{1900}{AC} = \tan 30^\circ$

$Ac = \frac{1900}{\tan 30^\circ}$

$\Rightarrow \quad AC = 1900\sqrt{3}$

Now BC = AC – BC

$\Rightarrow BC = 1900\sqrt{3} - \frac{1900}{\sqrt{3}} = 2193.93$

$\therefore$ Required distance $= 2194\,m.$

Problem 1. Solve the right triangle ABC, where $\gamma = 90^\circ$,

a =250, $\alpha = 42°25'$

Solution: As we know that: $\alpha + \beta + \gamma = 180^\circ$

$42°25' + \beta + 90^\circ = 180^\circ$

$\beta = 180^\circ - 132°25'$

$\beta = 47°35'$

$b = \frac{250}{\tan 42°25'}$

$b = \frac{250}{0.91365} = 273.63$

$c = \frac{250}{\sin 42°25'} = \frac{250}{0.67451} = 370.64$.

Problem 2. $a = 482, \quad \alpha = 35°36'$

Solution: $\tan \alpha = \frac{a}{b}$

$b = \frac{482}{\tan 35°36'}$

$b = 273.63$

$\sin \alpha = \frac{a}{c}$

$c = \frac{482}{\sin 35°36'}$

$c = 828$.

Problem 3. $a = 5, \quad c = 13$

Solution: By Pythagoras theorem: $c^2 = a^2 + b^2$

$b^2 = c^2 - a^2 = 13^2 - 5^2 = 169 - 25 = 144$

$b = \sqrt{144} \Rightarrow b = 12$

$\sin \alpha = \frac{a}{c}$

$\sin \alpha = \frac{5}{13} = 0.3846$

$\alpha = \sin^{-1} 0.3846$

$\alpha = 22°37'$

As we know that $\alpha + \beta + \gamma = 180°$

$22°37' + \beta + 90° = 180°$

$\beta = 180° - 125°36'$

$\beta = 67°23'$.

Problem 4. $b = 312, \quad \alpha = 23°42'$

Solution: As we know that $\alpha + \beta + \gamma = 180°$

$23°42' + \beta + 90° = 180°$

$\beta = 180° - 113°42'$

$\beta = 66°18'$

$\tan\alpha = \frac{a}{b}$

$a = \tan 23°42' \times 312$

$a = 136.96$

Now apply Pythagoras theorem $: c^2 = a^2 + b^2$

$c^2 = 136.96^2 + 312^2 = 18758.04 + 97344 = 116102.04$

$c = \sqrt{116102.04}$

$c = 340.74$.

Problem 5. $\mathbf{a} = 212, \ \boldsymbol{\beta} = 40°55'$

Solution: As we know that: $\alpha + \beta + \gamma = 180°$

$\alpha + 40°55' + 90° = 180°$

$\alpha = 180° - 130°55'$

$\alpha = 49°5'$

$\tan\alpha = \frac{a}{b}$

$b = \frac{212}{\tan 49°5'}$

$b = 183.74$

Now apply Pythagoras theorem $c^2 = a^2 + b^2$

$c^2 = 212^2 + 183.74^2$

$= 44944 + 33760.38 = 78704.38$

$c = \sqrt{78704.38}$

$c = 280.54$.

Problem 6. $\mathbf{c} = 232, \ \boldsymbol{\beta} = 52°46'$

Solution: As we know that: $\alpha + \beta + \gamma = 180°$

$\alpha + 52°46' + 90° = 180°$

$\alpha = 180° - 142°46'$

$\alpha = 37°14'$

$\sin\alpha = \frac{a}{c}$ => $a = 232 \times \sin 37°14'$ => $a = 140.37$

$\cos\alpha = \frac{b}{c}$ => $b = 232 \times \cos 37°14'$ => $b = 184.71$.

Problem 7. $a = 380, \quad c = 540$

Solution: Now apply Pythagoras theorem: $c^2 = a^2 + b^2$

$b^2 = c^2 - a^2$

$b^2 = 540^2 - 380^2 = 291600 - 144400$

$b^2 = 147200$

$b = \sqrt{147200}$

$b = 383.61$

$\sin\alpha = \frac{a}{c}$

$\sin\alpha = \frac{380}{540} = 0.7037$

$\alpha = \sin^{-1} 0.7037$

$\alpha = 44°43'$

As we know that: $\alpha + \beta + \gamma = 180°$

$44°43' + \beta + 90° = 180°$

$\beta = 180° - 134°43'$

$\beta = 45°16'$.

Problem 8. How far is a man from the foot of a tower 150m high, if the measure of the angle of elevation of its top as observed by him is 40°35'.

Solution: Let BC = Height of tower = 150 m,

AB = Distance from the foot of the tower = ?

In right ΔABC

$\tan 40°30' = \frac{BC}{AB}$

$$AB = \frac{150}{\tan 40°30'} = \frac{150}{0.8541}$$

$$AB = 175.63\text{m}.$$

Problem 9. The shadow of a building is 220m when the measure of the angle of elevation of the sun is 35°. Find the height of the building.

Solution:

Let $BC = h = \text{height of building},$

$AB = \text{lenght shadow of building} = 220\text{m}$

In right $\Delta\, ABC$

$$\tan 35° = \frac{BC}{AB}$$

$$BC = \tan 35° \times AB$$

$$BC = 0.70021 \times 220$$

$$BC = 154.05 \text{ m}.$$

Problem 10. A man 1.8 m tall observes that the angle of elevation of the top of a tree at distance of 12m from the man is 32°. What is the height of the tree?

Solution: Let $AD = \text{Height of man} = 1.8\text{m}$

$AB = \text{Distance between man and the tree} = 12\text{m}$

In right $\Delta\, DEC$

$$\tan 32° = \frac{EC}{DE}$$

$$EC = \tan 32° \times DE$$

$$EC = \tan 32° \times 12$$

$$EC = 7.4988$$

$$h - BC = EC$$

$$h - 1.8 = 7.4988$$

$$\boldsymbol{h} = 9.2988\,m.$$

Problem 11. On walking 300 m towards a tower in a horizontal line through its base, the measure of the angle of elevation of its top changes from 30° to 60°. Find the height of the tower.

Solution: In figure ΔBCD

$$\tan 60° = \frac{BC}{DC}$$

$$BC = \tan 60° \times DC$$

$$h = x(1.73) \qquad (1)$$

In ΔABC $\tan 30° = \frac{BC}{AC}$

$$BC = AC \times \tan 30°$$

$$h = (300 + x)(0.57).$$

From equation (1) and (2), we obtain:

$$x(1.73) = (300 + x)(0.57)$$

$$1.73x = 171 + 0.57x => 1.16x = 171$$

$$x = 147.41 \text{ m}$$

Now put these values into (1) to obtain:

$$h = 147.41(1.73)$$

$$h = 255.019.$$

Problem 12. The measure of the angle of elevation of a cliff is 25°. On walking 100m straight towards the cliff, the measure of the angle of elevation of the top is 48°.find the height of the cliff.

Solution: Let BC = height of cliff = h =?

$$AD = 100 \text{ m}$$

let $DB = X$

$$\tan 48° = \frac{BC}{BD}$$

$$BC = \tan 60° \times BD$$

$$h = x(1.11) \qquad (1)$$

In Δ ABC

$$\tan 25° = \frac{BC}{AB}$$

$$BC = AB \times \tan 25°$$

$$h = (100 + x)(0.466) \qquad (2)$$

from equation (1) & (2)

$$x(1.11) = (100 + x)(0.466)$$

$$1.11x = 46.6 + 0.466x \Rightarrow 0.644x = 46.6$$

$$x = 72.36m$$

Put this into equation n (1)

$$h = 72.36(1.11) \Rightarrow 80.37m\cdot$$

Problem 13. Take two points A and B 50 m apart and in the line with a tree, the measure of the angles of elevation of the top of the tree are 30° and 40° respectively. Find the height of the tree.

Solution: letCD = height of tree = h = ?

$$AB = 50m$$

$$AD = xm \Rightarrow DB = 50 - x.$$

From Δ ADC, $\tan 30° = \frac{h}{x}$

$$0.5774 = \frac{h}{x}$$

$$\Rightarrow x = \frac{h}{0.5774} = 1.7319h........(1)$$

From Δ DBC, $\tan 40° = \frac{h}{50 - x}$

$$0.8391 = \frac{h}{50 - x} \Rightarrow h = 41.95 - 0.8391x.......(2)$$

Put the value of x into (2)

$$h = 41.95 - 0.8391(1.7319h)$$

$$\Rightarrow 2.4532h = 41.95$$

$$h = \frac{41.95}{2.4532} = 1710m$$

Problem 14. Two men on the opposite sides of a tower observe that the measures of the angles of elevation of the tower separately as 15° and 25° respectively. If the height of the tower is 150 m. Find the distance between the observers.

Solution: CD = height of tower = 150 m

Let AD = xm   DB = ym

Let A and B be positions of the men.

AB=distance between two observer=?

In right $\Delta ABC, \tan 15° = \frac{150}{x}$

$$x = \frac{150}{0.2679} = 559.9 \text{ m}$$

In right, ΔBDC

$$\tan 25° = \frac{150}{y}$$

$$y = \frac{150}{0.4663} = 321.68$$

Thus $AB = x + y$

$$AB = 559.9 + 321.68$$

$$AB = 881.58 \text{ m}.$$

Problem 15. From a light-house, the angles of depression of two ships on opposite sides of the light-house are observed to be 30° and 45°. If the height of the light house is 300 m, find the distance between the ships of the line joining them that passes through the foot of the light-house.

Solution: CD is the light house. A and B are the positions of the ships.

$$\tan 45° = \frac{CD}{AD}$$

$$AD = \frac{CD}{\tan 45°} = \frac{300}{1} \Rightarrow AD = 300 \text{ m}$$

$$\tan 30° = \frac{CD}{BD}$$

$$BD = \frac{CD}{\tan 30°} = \frac{300}{0.577} \Rightarrow BD = 519.61 \text{ m}$$

Distance between ships $= AD + BD$

Distance between ships =300+519.61=819.6 m.

Problem 16. The measure of angle elevation of the top of a tower is 30° from a point on the ground. On retreating 100 m, the measure of the angle of elevation is found to be 15°. Find the height of the tower.

Solution: Let A and D be the position of the observer before and after retreating, and h be the height of the tower:

$$\tan 30° = \frac{BC}{AB}$$

$$AB = BC \times \cot 30°$$

$$\Rightarrow d = h(1.73)............(1)$$

$$\tan 15° = \frac{BC}{BD}$$

$$BD = BC \times \cot 15°$$

$$\Rightarrow (100 + d) = h(3.73)........(2)$$

Put these values into equation (1)

$$(100 + h(1.73)) = h(3.73)$$

$$100 = 3.73h - 1.73\ h$$

$$\Rightarrow h = \frac{100}{2} = 50 \text{ m}$$

Problem 17. From the top of a hill 200 m high, the angles of depression of the top and bottom of a tower are observed to be 30° and 60° respectively. Find the height of the tower.

Solution: AE = heigh tof tower = h = ?

BC = height of hill = 200 m

BD = AE = H => CD = BC – BD => 200 – H

Also AB = ED

From $\Delta ABC, \ \tan 60° = \frac{BC}{AB}$

$AB = \frac{200}{1.73}$

=> AB = 115.60 m

=> Thus, ED = 115.60

From $\Delta ABD, \ \tan 30° = \frac{CD}{ED}$

$0.577 = \frac{(200 - h)}{115.60}$

=> $200 - h = 0.577 \times 115.60$

=> h = 133.25 m

Height of the tower is 133.25 m.

Problem 18. A television antenna is on the roof of a building. From a point on the ground 36 m from the building, the angles of elevation of the top and the bottom of the antenna are 51° and 42° respectively. How tall is the antenna?

Solution: Height of roof is h and the total height with antenna is H

in $\Delta ABC, \ \tan 42° = \frac{BC}{AC}$

$BC = AC \times tant42°$

=> $h = 36 \times 0.9004 = 32.41m$

In $\Delta ACD, \ \tan 51° = \frac{CD}{AC}$

$CD = AC \times tant51°$

=> $H = 36 \times 1.23 = 44.45m$

Length of antenna $= H - h = 44.45 - 32.41 = 12.04$ m

Problem 19. A ladder 20 m long reaches a distance of 20 m from the top of a building. At the foot of the ladder the measure of the angle of elevation of the top of the building is 60°. Find the height of the building.

Solution: AD = length of ladder = 20 m

Let height of building = h =? =BC; CD=20m

Let $CAD = \alpha = 30°$,

$\Rightarrow BC = BD + DC$

$h = y + 20$

$\Rightarrow y = (h - 20)$

In ΔACD, $AD = CD = 20$ m

in right $\Delta BAD = 30°$.so $\sin 30° = \frac{BD}{AD}$

$\frac{1}{2} = \frac{y}{20}$

$\Rightarrow y = 10$ m

Substituting the value of y, we obtain

$h = y + 20$

$\Rightarrow h = 10 + 20 = 30$ m

Height of building = 30 m

Problem 20. A man standing of the bank of a canal observes that the measure of the angle of elevation of a tree is 60°. On retreating 40 m from the bank, he finds the angle of elevation of the tree as 30°. Find the height of the tree and the width of the canal.

Solution: let CD = height of tree = h = ?

BC = width of canal = x = ?

From ΔBCD, $\tan 60° = \frac{h}{x}$

$\sqrt{3} = \frac{h}{x}$

$=> h = x\sqrt{3}............(1)$

From ΔACD, $\tan 30° = \frac{CD}{AC}$

$\frac{1}{\sqrt{3}} = \frac{h}{40+x}$

$=> 40 + x = \sqrt{3}h..........(2)$

After substituting the value of h in (2), we obtain

$40 + x = \sqrt{3} \times x\sqrt{3}$

$=> 40 + x = 3x$

$=> x = 20$ m

Put value of x in (1)

$h = 20\sqrt{3} = 34.64$ m.

Problem 21. Take two buildings A and B as 100 m apart. The angle of elevation from the top of the building A to the top of building B is 20°. The angle of elevation from the base of the building B to the top of the building A is 50°. Find the height of building B.

Solution: BC = height of building = h = ?

AB = distance between building $A \& B = 100 m$

DE = AB = 100 m

FromΔDEC, $\tan 20° = \frac{EC}{DE}$

$0.3640 = \frac{EC}{100} => EC = 36.40$

FromΔABD, $\tan 50° = \frac{AD}{AB}$

$1.192 = \frac{AD}{100} => AD = 119.2 => BE = 119.2$

$h = BE + EC => 119.2 + 36.40 => h = 155.50$ m

SUMMARY

After completing this chapter, a student will be able to:

- Understand basic concepts of the right triangle.
- Identify right angle triangles.
- Understand basic concepts of the angle of elevation.
- Understand how an angle of elevation will be formed.
- Understand the basic concept of an angle of depression.
- Understand how angles of depression will be formed.
- Solve practical/mathematical problems related to the right angle triangle.

Review of common terms and basic formulas

Angle of elevation:

If the observer is below the object in their sight, the angle made by the lines of sight and the horizontal is called the "angle of elevation".

Angle of depression:

If the observer is above the object in their sight, the angle made by the lines of sight and the horizontal is called the "angle of depression".

CHAPTER 6

SOLUTION OF TRIANGLES

INTRODUCTION

A triangle has six important elements; three angles and three sides. In a triangle ABC, the measures of the three angles are usually denoted by α,β,γ and the measures of three sides opposite to these angles are denoted by a,b,c respectively.

If any three out of these six elements, with at least one side, are given, the remaining three elements can be determined. This process of finding the unknown elements is called the solution of triangles.

TYPES OF TRIANGLES

There are different types of triangles, based on the sides and angles.

a) Types based on the angle:

1. Acute Angle Triangle
2. Obtuse Angle Triangle
3. Right Angle Triangle

b) Types based sides

1. Equilateral Triangle
2. Isosceles Triangle
3. Scalene Triangle

OBLIQUE TRIANGLES

A triangle that does not have a right-angle is called an **oblique triangle.**

Solution of oblique triangles

We know that a triangle can be constructed if:

i) One side and two angles are given.

ii) Two sides and their included angle are given.

iii) Three sides are given.

In the same way, we can solve an oblique triangle if

- i) One side and two angles are known.
- ii) Two sides and their included angle are known.
- iii) Three sides are known.

Case I: When the measure of one side and two angles are given.

In this case, the law of sines can be applied.

Case II: When the measure of two sides and their included angle are given.

In this case, we can use any one of the following methods:

i) First law of cosine and then law of sines,

ii) First law of tangents and then law of sines.

We have learnt the methods of solving right triangles. However, in solving oblique triangle, we have to make use of the relations between the sides a, b, c and the angles α, β, γ, known as the law of cosines, law of sines and law of tangents.

Let us present these laws one by one, before solving oblique triangles.

THE LAW OF COSINE

In any triangle ABC, the square of any side is equal to the sum of the squares of the other two sides minus twice the product of these sides and the cosine of the included angle.

In any triangle *ABC,* with usual notations, prove that:

(i) $a^2 = b^2 + c^2 - 2bc\cos\alpha$

(ii) $b^2 = c^2 + a^2 - 2ca\cos\beta$

(iii) $c^2 = a^2 + b^2 - 2ab\cos\gamma$

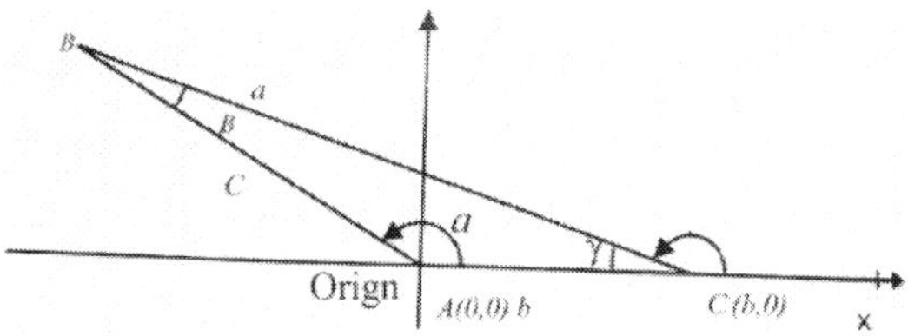

Figure 11

Proof: Let side $\overline{AC}$ of triangle *ABC* be along the positive direction of the x-axis with vertex A at the origin, then $<BAC$ will be in the standard position.

$\because AB = c \text{ and } m < BAC = a$

$\because$ coordinates of B are $(\cos\alpha, c\,a\sin\alpha)$

$\because$ AC = b and point C is on the x axis.

Coordinates of B are (b,0).

By the distance formula,

$|BC|^2 = (c\cos\alpha - b)^2 + (c\sin\alpha - 0)^2$

$$\Rightarrow \quad a^2 = c^2\cos^2\alpha + b^2 - 2bc\cos\alpha + c^2\sin^2\alpha \quad (\because \overline{BC} = a)$$

$$\Rightarrow \quad a^2 = c^2(\cos^2\alpha + \sin^2\alpha) + b^2 - 2bc\cos\alpha$$

$$\Rightarrow \quad a^2 = b^2 + c^2 - 2bc\cos\alpha.$$

In a similar way, we can prove that

$b^2 = c^2 + a^2 - 2ca\cos\beta$

$c^2 = a^2 + b^2 - 2ab\cos\gamma$

(i), (ii) and (iii) are called the law of cosines. They can also be expressed as:

$$\cos\alpha = \frac{b^2 + c^2 - a^2}{2bc}$$

$$\cos\beta = \frac{c^2 + a^2 - b^2}{2ca}$$

$$\cos\gamma = \frac{a^2 + b^2 - c^2}{2ab}.$$

Note: If ΔABC is right, then :

The law of cosines reduces to Pythagoras theorem i.e.,

If $\alpha = 90°$ *then* $b^2 + c^2 = a^2$

Or If $\beta = 90°$ *then* $c^2 + a^2 = b^2$

Or If $\gamma = 90°$ *then* $a^2 + b^2 = c^2$.

Problems related to the law of cosines:

Problem 1. $a = 56$ $c = 30$ $\beta = 35°$, find b.

Solution: Using the law of cosines.

$b^2 = c^2 + a^2 - 2\ ca\cos\beta$

$b^2 = 30^2 + 56^2 - 2(30)(56)\cos 35°$

$b^2 = 900 + 3136 - 2752.35$

$\Rightarrow b^2 = 1283.65$

$b = 35.82$.

Problem 2. $b = 25$ $c = 37$ $\alpha = 65°$, **find a**

Solution: Using the law of cosines.

$a^2 = b^2 + c^2 - 2bc\cos\alpha$

$a^2 = 25^2 + 37^2 - 2(25)(37)\cos 65°$

$a^2 = 625 + 1369 - 781.84$

$\Rightarrow a^2 = 1212.16$

$a = 34.82$

Problem 3. $b = 5$ $c = 8$ $\alpha = 60°$, **find a**

Solution: Using the law of cosines.

$a^2 = b^2 + c^2 - 2bc\cos\alpha$

$a^2 = 5^2 + 8^2 - 2(5)(8)\cos 60°$

$a^2 = 25 + 64 - 40$

$\Rightarrow a^2 = 49$

$a = 7$.

Problem 4. $a = 212 \quad c = 135 \quad \beta = 37°15'$, **find b**

Solution: Using the law of cosines.

$b^2 = c^2 + a^2 - 2ca\cos\beta$

$b^2 = 212^2 + 135^2 - 2(212)(135)\cos 37°15'$

$b^2 = 44944 + 18225 - 45563.15$

$\Rightarrow b^2 = 17605.85$

$b = 132.68$.

Problem 4. $a = 16 \quad c = 17 \quad \gamma = 25°$, **find c**

Solution: Using the law of cosines.

$c^2 = a^2 + b^2 - 2ab\cos\gamma$

$c^2 = 16^2 + 17^2 - 2(16)(17)\cos 25°$

$c^2 = 256 + 289 - 493.03$

$\Rightarrow c^2 = 51.97$

$c = 7.21$.

Problem 5. $a = 44 \quad c = 55 \quad \gamma = 114°$, **find c**

Solution: Using the law of cosines.

$c^2 = a^2 + b^2 - 2ab\cos\gamma$

$c^2 = 44^2 + 55^2 - 2(44)(55)\cos 114°$

$c^2 = 1936 + 3025 - 1968.61$

$\Rightarrow c^2 = 6929.61$

$c = 83.24$.

Problem 6. **a = 13 b = 10 c = 17, find α and β**

Solution: Using the law of cosines.

$$\cos\alpha = \frac{b^2 + c^2 - a^2}{2bc}$$

$$\cos\alpha = \frac{10^2 + 17^2 - 13^2}{2(10)(17)}$$

$$\cos\alpha = \frac{100 + 289 - 169}{340}$$

$$\Rightarrow \cos\alpha = \frac{220}{340}$$

$$\cos\alpha = 0.647$$

$$\Rightarrow \alpha = \cos^{-1} 0.647$$

$$\alpha = 49°40'47''$$

$$\cos\beta = \frac{a^2 + c^2 - b^2}{2ca}$$

$$\cos\beta = \frac{13^2 + 17^2 - 10^2}{2(13)(17)}$$

$$\cos\beta = \frac{139 + 289 - 100}{442}$$

$$\Rightarrow \cos\beta = \frac{358}{422}$$

$$\cos\beta = 0.809$$

$$\Rightarrow \beta = \cos^{-1} 0.809$$

$$\beta = 35°54'30''.$$

Problem 7. Three villages P, Q and R, are connected by straight roads. PQ is 6 km in length and measure QR is 9km in length. The angle between PQ and QR is 120°. Find the distance between P and Q.

Solution: Using the law of cosines.

$$q^2 = p^2 + r^2 - 2pr\cos Q$$

$$q^2 = 9^2 + 6^2 - 2(9)(6)\cos 120°$$

$$q^2 = 81 + 36 + 54$$

$$\Rightarrow q^2 = 171$$

$$q = 13.08 \text{ km.}$$

Problem 8. Two points A and B are at a distance 55 and 32 m respectively from a point P. The measure of angle between AP and BP is 37°. Find the distance between B and A.

Solution: $a = 32,\ b = 55\ P = 37°,$ find p

Using the law of cosines.

$$p^2 = a^2 + b^2 - 2ab\cos P$$

$$P^2 = 32^2 + 55^2 - 2(32)(55)\cos 37°$$

$$P^2 = 1024 + 3025 - 2811.2$$

$$\Rightarrow P^2 = 1237.8$$

$$p = 35.18\text{ m}.$$

Problem 9. Find the cosine of the smallest measure of the angle of a triangle with 12, 13 and 14 m as its sides.

Solution: Given that: $a = 12; b = 13; c = 14\ \alpha = ?$

$$\cos\alpha = \frac{b^2 + c^2 - a^2}{2bc}$$

$$\cos\alpha = \frac{(13)^2 + (14)^2 - (12)^2}{2(12)(13)}$$

$$\cos\alpha = \frac{169 + 196 - 144}{364}$$

$$\Rightarrow \cos\alpha = \frac{221}{364}$$

$$\cos\alpha = 0.607$$

$$\Rightarrow \alpha = \cos^{-1} 0.607$$

$$\alpha = 52°37'..$$

THE LAW OF SINES

If we know the two angles and the included side of a triangle (ASA), or if we know two angles and the side opposite one of the angles of a triangle (AAS), the size and shape of the triangle is determined. Therefore, we should be able to find the measures of the remaining sides.

In any triangle *ABC*, with usual notation, prove that:

$$\frac{a}{\sin\alpha}=\frac{b}{\sin\beta}=\frac{c}{\sin\gamma}$$

Proof: Let side $\overline{AC}$ of triangle *ABC* be along the positive direction of the x – axis with vertex A at origin, then $<BAC$ will be in the standard position.

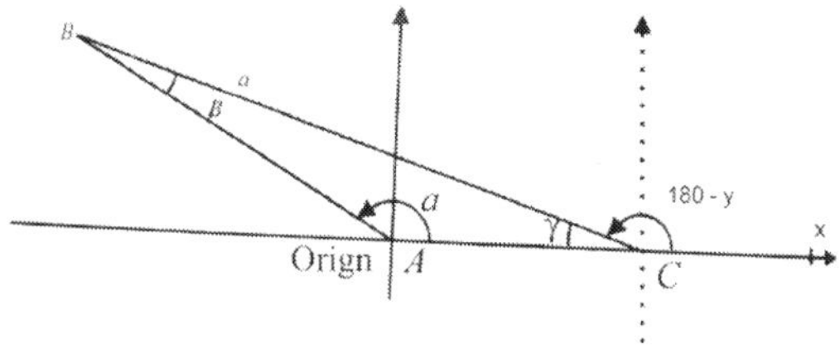

Figure 12

$\because AB = c$ and $m<BAC = \alpha$.

The coordinates of point B are (c cos α ,$c\sin\alpha$).

If the origin A is shifted to *C,* then $<BCX$ will be in the standard position,

$\because BC= a$ And $m<BCX = 180°-\gamma$

$\because$ The coordinates of *B* are $[a\cos(180°-\gamma), a\sin(180°-\gamma)]$.

In both cases. the-y coordinates of *B* remains the same:

$$\Rightarrow \frac{a}{\sin\alpha}=\frac{c}{\sin\gamma}.$$

In a similar way, with side $\overline{AB}$ along + ve x axis, we can prove that:

$$\frac{a}{\sin\alpha}=\frac{b}{\sin\beta}.$$

From (i) and (ii), we have $\frac{a}{\sin\alpha}=\frac{b}{\sin\beta}=\frac{c}{\sin\gamma}$.

This is called the law of sines.

SOME PROBLEMS RELATED TO THE LAW OF SINES: IN ANY TRIANGLE ABC IF:

Problem 1. a = 10 b = 15 β = 50°, find **α**

Solution: Using the law of sines.

$$\frac{a}{\sin\alpha} = \frac{b}{\sin\beta}$$

$$\Rightarrow \frac{10}{\sin\alpha} = \frac{15}{\sin 50°}$$

$$\sin\alpha = \frac{10 \times \sin 50°}{15}$$

$$\sin\alpha = 0.510$$

$$\alpha = \sin^{-1} 0.510$$

$$\Rightarrow \alpha = 30° 42' 36''.$$

Problem 2. a = 20 c = 32 γ = 70°, find α

Solution: Using the law of sines.

$$\frac{a}{\sin\alpha} = \frac{c}{\sin\gamma}$$

$$\Rightarrow \frac{20}{\sin\alpha} = \frac{32}{\sin 70°}$$

$$\sin\alpha = \frac{20 \times \sin 70°}{32}$$

$$\sin\alpha = 0.587$$

$$\alpha = \sin^{-1} 0.587$$

$$\Rightarrow á = 35°57'$$

Problem 3.a = 3 b = 7 = 85°, find **α**

Solution: Using the law of sins.

$$\frac{a}{\sin\alpha} = \frac{b}{\sin\beta}$$

$$\Rightarrow \frac{3}{\sin\alpha} = \frac{7}{\sin 85°}$$

$$\sin\alpha = \frac{3 \times \sin 85°}{7}$$

$$\sin\alpha = 0.426$$

$$\alpha = \sin^{-1} 0.426$$

$$\Rightarrow \alpha = 25°16'24''$$

Problem 4. a = 5 c = 6 α = 45°, **find**

Solution: Using the law of sines.

$$\frac{a}{\sin\alpha} = \frac{c}{\sin\gamma}$$

$$\Rightarrow \frac{5}{\sin 45^\circ} = \frac{6}{\sin \gamma}$$

$$\sin \gamma = \frac{6 \times \sin 45^\circ}{5}$$

$$\sin \gamma = 0.846$$

$$\gamma = \sin^{-1} 0.846$$

$$\Rightarrow \gamma = 58^\circ 3'.$$

Problem 5. a = $20\sqrt{3}$ **γ =75° = 60°, find c**

Solution: Using the law of sines.

$$\frac{a}{\sin \alpha} = \frac{c}{\sin \gamma}$$

$$\Rightarrow \frac{20\sqrt{3}}{\sin 60^\circ} = \frac{c}{\sin 75^\circ}$$

$$c = \frac{20\sqrt{3} \times \sin 75^\circ}{\sin 60^\circ}$$

$$\Rightarrow c = 31.06$$

Problem 6. a = 211.3 β = 48°16' **γ** = 71°38', **find b**

Solution: As we know that: $\alpha + \beta + \gamma = 180^\circ$

$$\alpha + 48^\circ 16' + 71^\circ 38' = 180^\circ$$

$$\alpha = 180^\circ - 119^\circ 54' = 60^\circ 6'$$

Using the law of sines.

$$\frac{a}{\sin \alpha} = \frac{b}{\sin \beta}$$

$$\Rightarrow \frac{211.3}{\sin 60^\circ 6'} = \frac{b}{\sin 48^\circ 16'}$$

$$b = \frac{211.3 \times \sin 48^\circ 16'}{\sin 60^\circ 6'}$$

$$\Rightarrow b = 181.89.$$

Problem 7. a = 18 α= 47° β = 102°, find c

Solution: As we know that: $\alpha + \beta + \gamma = 180^\circ$

$$47^\circ + 102^\circ + \gamma = 180^\circ$$

$$\gamma = 180^\circ - 149^\circ = 31^\circ$$

Using the law of sines.

$$\frac{c}{\sin \gamma} = \frac{a}{\sin \alpha}$$

$$\Rightarrow \frac{c}{\sin 31°} = \frac{18}{\sin 47°}$$

$$c = \frac{18 \times \sin 31°}{\sin 47°}$$

$$\Rightarrow c = 12.68.$$

Problem 8. a = 475 β = 72°15’ **γ** = 43°30’, **find b**

Solution: As we know that: $\alpha + \beta + \gamma = 180°$

$$\alpha + 72°15' + 43°30' = 180°$$

$$\alpha = 180° - 115°45' \Rightarrow \alpha = 64°15'$$

Using the law of sines.

$$\frac{a}{\sin \alpha} = \frac{b}{\sin \beta}$$

$$\Rightarrow \frac{2475}{\sin 64°15'} = \frac{b}{\sin 72°15'}$$

$$b = \frac{211.3 \times \sin 72°15'}{\sin 64°15}$$

$$\Rightarrow b = 502.3$$

Problem 9. b = 82 β = 57° **γ** = 78°, find **a**

Solution: As we know that: $\alpha + \beta + \gamma = 180°$

$$\alpha + 57° + 78° = 180°$$

$$\alpha = 180° - 135°$$

$$\Rightarrow \alpha = 45°$$

Using the law of sines.

$$\frac{a}{\sin \alpha} = \frac{b}{\sin \beta}$$

$$\Rightarrow \frac{a}{\sin 45°} = \frac{82}{\sin 57°}$$

$$a = \frac{82 \times \sin 45°}{\sin 57°}$$

$$\Rightarrow a = 69.13$$

Problem 10. **α** = 60° **β** = 45°, find the ratio of b to c.

Solution: As we know that: $\alpha + \beta + \gamma = 180°$

$$60° + 45° + \gamma = 180°$$

$\gamma = 180° - 105°$

$\Rightarrow \gamma = 75°$

Using the law of sines.

$\frac{B}{\sin\beta} = \frac{c}{\sin\gamma}$

$\Rightarrow \frac{b}{\sin 45°} = \frac{c}{\sin 60°}$

$\frac{b}{c} = \frac{\sin 45°}{\sin 60°}$

$\Rightarrow b : c = 0.732\,..$

Problem 11. Two shore batteries, positioned at A and B, are 840 m apart and are firing at target C. The measure of ABC is 80° and the measure of angle BAC is 70°. Find the measure of distance AC and BC.

Solution: As we know that: A+B+C=180°

$70° + 80° + C = 180°$

$C = 180° - 150°$

$\Rightarrow C = 30°$

Using the law of sines.

$\frac{c}{\sin C} = \frac{a}{\sin A}$; $\frac{b}{\sin B} = \frac{c}{\sin C}$

$\frac{840}{\sin 30°} = \frac{a}{\sin 70°}$; $\frac{b}{\sin 80°} = \frac{840}{\sin 30°}$

$a = \frac{840 \times \sin 70°}{\sin 30°}$; $b = \frac{840 \times \sin 80°}{\sin 30°}$

$a = 1578.68$ m ; $b = 1654.45$ m.

THE LAW OF TANGENTS

In any triangle *ABC,* with usual notations, prove that:

i) $$\frac{a-b}{a+b} = \frac{\tan\frac{a-b}{2}}{\tan\frac{a+b}{2}}$$

ii) $\frac{b-c}{b+c}=\frac{\tan\frac{\beta-\gamma}{2}}{\tan\frac{\beta+\gamma}{2}}$

iii) $\frac{c-a}{c+a}=\frac{\tan\frac{\gamma-a}{2}}{\tan\frac{\gamma+a}{2}}$

Proof: We know that by the law of sines:

$$\frac{a}{\sin\alpha}=\frac{b}{\sin\beta}$$

$$\Rightarrow \frac{a}{b}=\frac{\sin\alpha}{\sin\beta}$$

By componendo and dividendo,

$$\frac{a-b}{a+b}=\frac{\sin a-\ \sin\beta}{\sin a+\ \sin\beta}=\frac{2\cos\frac{a+\beta}{2}\ \sin\frac{\alpha-\beta}{2}}{2\ \sin\frac{a+\beta}{2}\ \cos\ \frac{\alpha-\beta}{2}}$$

$$\Rightarrow \frac{a-b}{a+b}=\frac{tan\frac{\alpha-\beta}{2}}{tan\frac{\alpha+\beta}{2}}$$

Similarly, we can prove that:

x $\frac{b-c}{b+c}=\frac{tan\frac{\beta-\gamma}{2}}{\tan\frac{\beta+\gamma}{2}}$ (ii)

. and $\frac{c-a}{c+b}=\frac{\tan(\frac{\gamma-a}{2})}{\tan\ (\frac{\gamma+\alpha}{2})}$ (iii)

$\Rightarrow$ (i), (ii) and (iii) are called Law of Tangents.

Example: The two sides of a triangle have a ratio 3:2 and have an angle of 57°.Find the remaining two angles

Solution: Let $a:b=3:2$

i.e. $\frac{a}{b}=\frac{3}{2}$

$\Rightarrow a=\frac{3}{2}b$

and $\gamma=57°$

Since $\alpha+\beta+\gamma=180°$

$\Rightarrow \alpha + \beta = 180^\circ - \gamma$

$\Rightarrow \alpha + \beta = 180^\circ - 57^\circ$

$\Rightarrow \alpha + \beta = 123^\circ$ (1)

Now, by law of tangents

$$\frac{a-b}{a+b} = \frac{\tan\left(\frac{\alpha-\beta}{2}\right)}{\tan\left(\frac{\alpha+\beta}{2}\right)}$$

$$\Rightarrow \frac{\frac{3}{2}b-b}{\frac{3}{2}b+b} = \frac{\tan\left(\frac{\alpha-\beta}{2}\right)}{\tan\left(\frac{123^0}{2}\right)}$$

$$\Rightarrow \frac{\frac{1}{2}b}{\frac{5}{2}b} = \frac{\tan\left(\frac{\alpha-\beta}{2}\right)}{\tan\left(61^0 30'\right)}$$

$$\Rightarrow \frac{\frac{1}{2}}{\frac{5}{2}} = \frac{\tan\left(\frac{\alpha-\beta}{2}\right)}{1.8418}$$

$$\Rightarrow \frac{1}{5}(1.8418) = \tan\left(\frac{\alpha-\beta}{2}\right)$$

$$\Rightarrow \tan\left(\frac{\alpha-\beta}{2}\right) = 0.3684$$

$\left(\frac{\alpha-\beta}{2}\right) = \tan^{-1}(0.3684) = 20^\circ 13'$

$\alpha - \beta = 40^\circ 26'$ (2)

Adding (1) and (2), we obtain

$2\alpha = 163^\circ 27'$

$\alpha = 81^\circ 44'$

Now, put the value of α into equation (1) to obtain

$81^\circ 44' + \beta = 123^\circ$

$\Rightarrow \beta = 123^\circ - 81^\circ 44'$

$\Rightarrow \beta = 41^\circ 16'$.

Example: Using the law of tangents, solve the triangle $\boldsymbol{a} = 36.21$, $\boldsymbol{b} = 42.09$, $\gamma = 44^\circ 29'$

Solution: $\alpha + \beta + \gamma = 180^\circ$

$\Rightarrow \alpha + \beta = 180^\circ - \gamma$

$\Rightarrow \alpha + \beta = 180^\circ - 44^\circ 29'$

$\Rightarrow \alpha + \beta = 135^\circ 31'$ (1)

Now, by the law of tangents

$$\frac{a-b}{a+b} = \frac{\tan\left(\frac{\alpha-\beta}{2}\right)}{\tan\left(\frac{\alpha+\beta}{2}\right)}$$

$$\Rightarrow \frac{36.21-42.09}{36.21+42.09} = \frac{\tan\left(\frac{\alpha-\beta}{2}\right)}{\tan\left(\frac{135^\circ 31'}{2}\right)}$$

$$\Rightarrow \frac{-5.88}{78.3} = \frac{\tan\left(\frac{\alpha-\beta}{2}\right)}{\tan\left(67^\circ 45'\right)}$$

$$\Rightarrow -0.0751 = \frac{\tan\left(\frac{\alpha-\beta}{2}\right)}{2.4443}$$

$$\tan\left(\frac{\alpha-\beta}{2}\right) = -0.0751(2.4443)$$

$$\tan\left(\frac{\alpha-\beta}{2}\right) = -0.1836$$

$$\frac{\alpha-\beta}{2} = \tan^{-1}(-0.1836)$$

$$\frac{\alpha-\beta}{2} = -10^\circ 24'$$

$\Rightarrow \alpha - \beta = -20^\circ 48'$. (2)

Adding (1) & (2), we obtain

$2\alpha = 114^{\circ}43'$

$\alpha = 57^{\circ}22'$

Put value of α in (1)

$57^{\circ}22' + \beta = 135^{\circ}22'$

$\Rightarrow \beta = 78^{\circ}9'$

Now by the law of sines:

$$\frac{c}{\sin\gamma} = \frac{a}{\sin\alpha}$$

$$\Rightarrow \frac{c}{\sin 44^{\circ}29'} = \frac{36.21}{\sin 57^{\circ}22'}$$

$$\Rightarrow \quad c = \frac{36.21}{\sin 57^{\circ}22'} .\sin 44^{\circ}29'$$

$$\Rightarrow \quad c = 30.13.$$

Example: Using the law of tangents, solve the triangle $b = 61, \ c = 32, \ \alpha = 59^{\circ}30'$

Solution: Since $\alpha + \beta + \gamma = 180^{\circ}$

$\Rightarrow \ \beta + \gamma = 180^{\circ} - \alpha$

$\Rightarrow \ \beta + \gamma = 180 - 59^{\circ}30'$ s

$\Rightarrow \ \beta + \gamma = 120^{\circ}30'$ (1)

By law of tangent

$$\frac{b-c}{b+c} = \frac{\tan\left(\frac{\beta-\gamma}{2}\right)}{\tan\left(\frac{\beta+\gamma}{2}\right)}$$

$$\Rightarrow \frac{61-32}{61+32} = \frac{\tan\left(\frac{\beta-\gamma}{2}\right)}{\tan\left(\frac{120^{\circ}30'}{2}\right)}$$

$$\Rightarrow \frac{29}{93} = \frac{\tan\left(\frac{\beta-\gamma}{2}\right)}{\tan\left(60^{\circ}15'\right)}$$

$$\Rightarrow 0.3118 = \frac{\tan\left(\frac{\beta-\gamma}{2}\right)}{1.7496}$$

$$\Rightarrow \tan\left(\frac{\beta-\gamma}{2}\right) = 0.3118(1.7496)$$

$\tan\left(\frac{\beta-\gamma}{2}\right)=0.5455$

$\frac{\beta-\gamma}{2}=\tan^{-1}(0.5455)$

$\frac{\beta-\gamma}{2}=28^{\circ}37'$

$\Rightarrow \beta-\gamma=57^{\circ}14'$. (2)

Adding (1) & (2), we obtain

$2\beta=177^{\circ}44'$

$\beta=88^{\circ}52'$.

Substituting the value of α into (1)

$88^{\circ}52'+\gamma=120^{\circ}30'$

$\Rightarrow \gamma=31^{\circ}38'$

Now, by law of sines:

$\frac{c}{\sin\gamma}=\frac{a}{\sin\alpha}$

$\Rightarrow \frac{36}{\sin 31^{0}38'}=\frac{a}{\sin 59^{0}30'}$

$\Rightarrow \quad a=\frac{36}{\sin 59^{0}30'}.\sin 31^{\circ}38'$

$\Rightarrow \quad c=22.04$.

HALF ANGLE FORMULAS

We shall now prove the so-called half angle formulas, with the help of the law of cosines:

a) The sine of the half the angle in terms of the sides

In any triangle *ABC,* prove that:

(i) $\sin\frac{\alpha}{2}=\sqrt{\frac{(s-b)(s-c)}{bc}}$

(ii) $\sin\frac{\beta}{2}=\sqrt{\frac{(s-c)(s-a)}{ca}}$ where $2s = a + b + c$

(iii) $\sin\frac{\gamma}{2}=\sqrt{\frac{(s-a)(s-b)}{ab}}$.

Proof: We know that

$$2\sin^2\frac{\alpha}{2}=1-\cos\alpha$$

$$\therefore\quad 2\sin^2\frac{\alpha}{2}=1-\frac{b^2+c^2-a^2}{2bc}$$

$$\left\{\because \text{á}\cos=\frac{b^2+c^2-a^2}{2bc}\right.$$

$$\therefore\quad 2\sin^2\frac{\alpha}{2}=\frac{a^2-\left(b^2+c^2-2bc\right)}{2bc}=\frac{a^2-(b-c)^2}{2bc}$$

$$\therefore\quad \sin^2\frac{\alpha}{2}=\frac{(a^2+b-c)(a-b+c)}{4bc}\qquad\{\therefore a+b+c=2s\}$$

$$\text{Hence:}\quad \sin^2\frac{\alpha}{2}=\frac{2(s-b)(s-c)}{bc}$$

$$\left\{\begin{array}{l}\therefore\quad \alpha\ \text{is the measure of}\\ \text{an angle of}\ \Delta ABC\\ \therefore\ \frac{\alpha}{2}<90^\circ\Rightarrow\sin\frac{\alpha}{2}=+\text{ve}\end{array}\right.$$

In a similarly way, we can prove that

$\sin\frac{\beta}{2}=\sqrt{\frac{(s-c)(s-c)}{ca}}$ and $\sin\frac{\gamma}{2}=\sqrt{\frac{(s-a)(s-b)}{ab}}$.

b) The cosine of the half angle in term of the sides

In any triangle ABC, with usual notation, prove that:

$$\cos\frac{\alpha}{2}=\sqrt{\frac{s(s-a)}{bc}}$$

$$\cos\frac{\beta}{2}=\sqrt{\frac{s(s-b)}{ac}}$$

$$\cos\frac{\gamma}{2}=\sqrt{\frac{s(s-c)}{ab}}$$

Proof: We know that:

$$2\cos^2\frac{\alpha}{2}=1+cos\alpha$$

$$2\cos^2\frac{\alpha}{2}=1+\left(\frac{b^2+c^2-a^2}{2bc}\right)$$

$$2\cos^2\frac{\alpha}{2}=\frac{2bc+b^2+c^2-a^2}{2bc}$$

$$2\cos^2\frac{\alpha}{2}=\frac{(b+c)^2-a^2}{2bc}$$

$$2\cos^2\frac{\alpha}{2}=\frac{(b+c+a)(b+c-a)}{2bc}$$

$$\cos^2\frac{\alpha}{2}=\frac{2s(2s-2a)}{4bc}$$

$$\cos^2\frac{\alpha}{2}=\frac{s(s-a)}{bc}$$

$$\cos\frac{\alpha}{2}=\sqrt{\frac{s(s-a)}{bc}}.$$

In a similar way, we can prove that:

$$\cos\frac{\beta}{2}=\sqrt{\frac{s(s-b)}{ca}}$$

$$\cos\frac{\gamma}{2}=\sqrt{\frac{s(s-c)}{ab}}.$$

c) The tangent of the half angle in terms of the sides

In any triangle ABC, with the usual notation, prove that:

$$\tan\frac{\alpha}{2}=\sqrt{\frac{(s-b)(s-a)}{s(s-a)}}$$

$$\tan\frac{\beta}{2}=\sqrt{\frac{(s-c)(s-a)}{s(s-b)}}$$

$$\tan\frac{\gamma}{2}=\sqrt{\frac{(s-a)(s-b)}{s(s-c)}}.$$

Proof: We know that:

$$\sin\frac{\alpha}{2}=\sqrt{\frac{(s-b)(s-c)}{bc}}$$

and $\cos\frac{\alpha}{2}=\sqrt{\frac{s(s-a)}{bc}}$

$$\tan\frac{\alpha}{2}=\frac{\sin\frac{\alpha}{2}}{\cos\frac{\alpha}{2}}=\frac{\sqrt{\frac{(s-b)(s-c)}{bc}}}{\sqrt{\frac{s(s-a)}{bc}}}$$

$$\tan\frac{\alpha}{2}=\sqrt{\frac{(s-b)(s-c)}{s(s-a)}}.$$

In a similar way, we can prove that:

$$\tan\frac{\beta}{2}=\sqrt{\frac{(s-c)(s-a)}{s(s-b)}}$$

$$\tan\frac{\gamma}{2}=\sqrt{\frac{(s-a)(s-b)}{s(s-c)}}.$$

Example 3: Two forces of 20 N and 15 N, inclined at an angle, are applied at a point on a body. If these forces are represented by two adjacent of a parallelogram, then their resultant is represented by its diagonal. Find the resultant forces and also the angle which the resultant makes with the force of 20 N.

Solution:

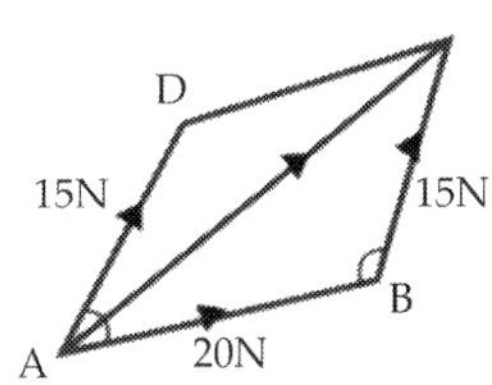

Let ABCD be a ll', such that $\overrightarrow{AB}$ represent 20 Newtons,

$\overrightarrow{AD}$ represent 15 Newtons,

and m =

ABCD is ll"

$$\left\{\begin{array}{c} \overline{BC} = \overline{AD} \ = 15\,N \\ m > ABC = 180° = -M > BAD = 180° - 45° = 135° \end{array}\right\}.$$

By the law of cosines:

$$\left(\overline{AC}\right)^2 = \left(\overline{AB}\right)^2 + \left(\overline{BC}\right)^2 - 2\times\left(\overline{AB}\right)^2 \times \left(\overline{BC}\right)^2 \times \cos 135°$$

$$= (20)^2 + (15)^2 - 2 \times 20 \times 15 \times \left(\frac{-1}{\sqrt{2}}\right)$$

$$= 400 + 225 + 424.2$$

=1049.2.

$$\therefore \qquad \overline{AC} = \sqrt{1049.2} = 32.4\,N$$

By the law of sines:

$$\frac{\overline{BC}}{\sin m\angle BAC} = \frac{\overline{AC}}{\sin 135°}$$

$$\therefore \qquad \sin m\angle BAC = \frac{\overline{BC}\, x \sin 135°}{\overline{AC}} = \frac{15 \text{ x} 0.707}{32.4} = 0.3274$$

$$\therefore \qquad m > BAC = 19°6'.$$

Example: Two forces of 40 N and 30 N are represented by AB and BC, inclined at an angle of 147° 25'. Find AC, the resultant of AB and BC.

Solution: Since it is given that $\overline{AB} = c = 40\,N$

$$\overline{BC} = a = 30\,N$$

$$\beta = 147^0\,25'$$

$$\overline{AC} = b = ?$$

Now by law of cosines:

$$b^2 = c^2 + a^2 - 2ca\cos\beta$$

$$b^2 = (40)^2 + (30)^2 - 2ca\cos(147^0\,25')$$

$$b^2 = 1600 + 900 - 2(40)(30)(-0.8426)$$

$b^2 = 4522.26$

$\Rightarrow b = 67.248$

i.e., $\overline{AC} = 67.248\,N$.

Example: Solve the triangle ABC using the law of cosines when a = 7, b=3 , c = 5.

Solution: We know that

$$\cos\alpha = \frac{b^2+c^2+a^2}{2bc}$$

$$\cos\alpha = \frac{9+25-19}{30} = -\frac{15}{30} = -\frac{1}{2}$$

$\alpha = 120°$

$$\cos\beta = \frac{c^2+a^2+ba^2}{2ca} = \frac{25+49-9}{70} = \frac{65}{70} = 0.9286$$

$\therefore\ \beta = 21°47$

and $\gamma = 180° - (\alpha+\beta) = 180° - (120 + 21°47) = 38°13$.

Example: Solve the triangle ABC using the half angle formula, when a = 283, b=317, c= 428.

Solution: 2s = a+ b+c = 283 + 317 + 428 = 1028

$s = 514$

$s - a = 514 - 283 = 231$

$s - b = 514 - 317 = 197$

$s - c = 514 - 428 = 96$.

Now, $\tan\frac{\alpha}{2} = \frac{\sqrt{(s-c)(s-a)}}{s(s-b)} = \frac{\sqrt{197x86}}{514x231} = 0.3777$

$\frac{\alpha}{2} = 20°42 \Rightarrow \alpha\, 41°24$

and $\tan\frac{\beta}{2} = \frac{\sqrt{(s-c)(s-a)}}{s(s-b)} = \frac{\sqrt{86x231}}{514x197} = 0.442$

$\therefore \frac{\beta}{2} = 23°53 \Rightarrow \beta = 47°46$

$\gamma = 180° - (\alpha + \beta) = 180° - (41°24 + 47°46) =$

Solve the triangle ABC with the given data:

Example: $c = 4, \quad \alpha = 70°, \quad \gamma = 42°$

Solution: We know that: $\alpha + \beta + \gamma = 180°$

$70° + \beta + 42° = 180°$

$\beta = 180° - 112° = 68°$

Using law of sines:

$\frac{b}{\sin\beta} = \frac{c}{\sin\gamma} \quad ; \quad \frac{a}{\sin\alpha} = \frac{c}{\sin\gamma}$

$\frac{b}{\sin 68°} = \frac{4}{\sin 42°} \quad ; \quad \frac{a}{\sin 70°} = \frac{4}{\sin 42°}$

$b = \frac{4 \times \sin 68°}{\sin 42°} \quad ; \quad a = \frac{4 \times \sin 70°}{\sin 42°}$

$b = 5.54 \quad ; \quad a = 5.62$

Example: $a = 464°, \quad \beta = 102°, \quad \gamma = 23°$

Solution: We know that: $\alpha + \beta + \gamma = 180°$

$\alpha + 102° + 23° = 180°$

$\alpha = 180° - 125° = 55°$

Using the law of sines:

$\frac{b}{\sin\beta} = \frac{a}{\sin\alpha} \quad ; \quad \frac{a}{\sin\alpha} = \frac{c}{\sin\gamma}$

$\frac{b}{\sin 102°} = \frac{464}{\sin 55°} \quad ; \quad \frac{464}{\sin 55°} = \frac{c}{\sin 23°}$

$b = \frac{464 \times \sin 102°}{\sin 55°} \quad ; \quad c = \frac{464 \times \sin 23°}{\sin 55°}$

$b = 554.06 \quad ; \quad c = 221.32.$

Example: $b = 85°$ $\beta = 57°15'$ $\gamma = 78°18'$

Solution: We know that: $\alpha + \beta + \gamma = 180°$

$$\alpha + 57°15' + 78°18' = 180°$$

$$\alpha = 180° - 135°33' \Rightarrow \alpha = 44°27'.$$

Using the law of sines:

$$\frac{b}{\sin\beta} = \frac{a}{\sin\alpha} \quad ; \quad \frac{a}{\sin\alpha} = \frac{c}{\sin\gamma}$$

$$\frac{85}{\sin 57°15'} = \frac{a}{\sin 44°27'} \quad ; \quad \frac{85}{\sin 44°27'} = \frac{c}{\sin 78°18'}$$

$$a = \frac{85 \times \sin 44°27'}{\sin 57°15'} \quad ; \quad c = \frac{85 \times \sin 78°18'}{\sin 44°27'}$$

$$a = 70.77 \quad ; \quad c = 98.96.$$

Example: $b = 56.8$ $\alpha = 79°31'$ $\beta = 44°24'$

Solution: We know that: $\alpha + \beta + \gamma = 180°$

$$79°31' + 44°24' + \gamma = 180°$$

$$\gamma = 180° - 123°55'$$

$$\gamma = 56°5'.$$

Using the law of sines:

$$\frac{b}{\sin\beta} = \frac{a}{\sin\alpha} \quad ; \quad \frac{b}{\sin\beta} = \frac{c}{\sin\gamma}$$

$$\frac{56.8}{\sin 44°24'} = \frac{a}{\sin 79°31'} \quad ; \quad \frac{56.8}{\sin 44°24'} = \frac{c}{\sin 56°5'}$$

$$a = \frac{56.8 \times \sin 79°31'}{\sin 44°24'} \quad ; \quad c = \frac{56.8 \times \sin 56°5'}{\sin 44°24'}$$

$$a = 79.82 \quad ; \quad c = 67.37.$$

Example: $b = 34.57, \alpha = 62°11', \beta = 63°22'$

Solution: We know that

$$\alpha + \beta + \gamma = 180°$$

$$62°11' + 63°22' + \gamma = 180°$$

$$\gamma = 180° - 125°33' = 54°27'.$$

By using laws of sine:

$$\frac{b}{\sin\beta}=\frac{a}{\sin\alpha} \quad ; \quad \frac{b}{\sin\beta}=\frac{c}{\sin\gamma}$$

$$\frac{34.57}{\sin 63°22'}=\frac{a}{\sin 62°11'} \quad ; \quad \frac{34.57}{\sin 63°22'}=\frac{c}{\sin 54°27'}$$

$$a=\frac{34.57\times \sin 62°11'}{\sin 63°22'} \quad ; \quad c=\frac{34.57\times \sin 54°27'}{\sin 63°22'}$$

$$a=34.20 \quad ; \quad c=31.47$$

Example: Find the angle of largest measure in the triangle ABC, where:

$\mathbf{a}=224$ $\mathbf{b}=380$ $\mathbf{c}=340$, $\boldsymbol{\beta}$**=?**

Solution: $\cos\boldsymbol{\beta}=\dfrac{\mathbf{a}^2+\mathbf{c}^2-\mathbf{b}^2}{\mathbf{2ac}}$

$$\cos\beta=\frac{(224)^2+(380)^2-(340)^2}{2(224)(340)}$$

$$\cos\beta=\frac{50176+115600-144400}{152320}$$

$$\cos\beta=\frac{21376}{152320}$$

$$\cos\beta=0.140$$

$$\beta=\cos^{-1}0.140$$

$$\boldsymbol{\beta}=81°55'57''.$$

Example: $\mathbf{a}=374$, $\mathbf{b}=514$, $\mathbf{c}=425$, $\boldsymbol{\beta}$**=?**

Solution: $\cos\beta=\dfrac{a^2+c^2-b^2}{2ac}$

$$\cos\beta=\frac{(374)^2+(514)^2-(425)^2}{2(374)(514)}$$

$$\cos\beta=\frac{139876+180625-264196}{317900}$$

$$\cos\beta=\frac{56305}{317900}$$

$\cos\beta = 0.177$

$\beta = \cos^{-1} 0.177$

$\beta = 79°47'53"$.

Example: Solve the triangle ABC, where:

a = 74, **b** = 52, **c** = 47

Solution: $\cos\alpha = \dfrac{b^2 + c^2 - a^2}{2bc}$

$$\cos\alpha = \frac{(52)^2 + (47)^2 - (74)^2}{2(52)(47)}$$

$$\cos\alpha = \frac{2704 + 2209 - 5476}{4888}$$

$$\cos\alpha = \frac{-563}{4888}$$

$\cos\alpha = -0.115$

$\alpha = \cos^{-1}(-0.115)$

$\alpha = 96°36'50"$

$$\cos\beta = \frac{a^2 + c^2 - b^2}{2ac}$$

$$\cos\beta = \frac{(74)^2 + (47)^2 - (52)^2}{2(74)(47)}$$

$$\cos\beta = \frac{5476 + 2209 - 2704}{6956}$$

$$\cos\beta = \frac{4981}{6956}$$

$\cos\beta = 0.716$

$\beta = \cos^{-1} 0.716$

$\beta = 44°16'78"$

$$\cos\gamma = \frac{a^2 + b^2 - c^2}{2ab}$$

$$\cos\gamma = \frac{(74)^2 + (52)^2 - (47)^2}{2(74)(52)}$$

$$\cos\gamma = \frac{5476 + 2704 - 2209}{7696}$$

$$\cos\gamma = \frac{5971}{9696}$$

$$\cos\gamma = 0.775$$

$$\gamma = \cos^{-1}(0.775)$$

$$\gamma = 39°7'1''$$

Example: Solve the triangle ABC, where:

a = 7 **b** = 9 **c** = 7

Solution: $\cos\alpha = \dfrac{b^2 + c^2 - a^2}{2bc}$

$$\cos\alpha = \frac{(9)^2 + (7)^2 - (7)^2}{2(7)(7)}$$

$$\cos\alpha = \frac{81 + 49 - 49}{126}$$

$$\cos\alpha = \frac{81}{126}$$

$$\cos\alpha = 0.642$$

$$\alpha = \cos^{-1} 0.642$$

$$\alpha = 49°59'$$

$$\cos\beta = \frac{a^2 + c^2 - b^2}{2ac}$$

$$\cos\beta = \frac{(7)^2 + (7)^2 - (9)^2}{2(7)(7)}$$

$$\cos\beta = \frac{49 + 49 - 81}{98}$$

$$\cos\beta = \frac{17}{986}$$

$$\cos\beta = 0.173$$

$$\beta = \cos^{-1} 0.173$$

$$\beta = 80°$$

$$\cos\gamma = \frac{a^2 + b^2 - c^2}{2ab}$$

$$\cos\gamma = \frac{(7)^2 + (9)^2 - (7)^2}{2(7)(9)}$$

$$\cos\gamma = \frac{49 + 81 - 49}{126}$$

$$\cos\gamma = \frac{81}{126}$$

$$\cos\gamma = 0.642$$

$$\gamma = \cos^{-1} 0.642$$

$$\gamma = 49°59'.$$

Example: Solve the triangle ABC, where:

$\mathbf{a} = 2.3, \mathbf{b} = 1.5, \mathbf{c} = 2.7$

Solution: $\cos\alpha = \dfrac{b^2 + c^2 - a^2}{2bc}$

$$\cos\alpha = \frac{(1.5)^2 + (2.7)^2 - (2.3)^2}{2(1.5)(2.7)}$$

$$\cos\alpha = \frac{2.25 + 7.29 - 5.29}{8.1}$$

$$\cos\alpha = \frac{4.25}{8.1}$$

$$\cos\alpha = 0.524$$

$$\alpha = \cos^{-1} 0.524$$

$\alpha = 58°21'$

$$\cos\beta = \frac{a^2 + c^2 - b^2}{2ac}$$

$$\cos\beta = \frac{(2.3)^2 + 2.7^2 - (1.5)^2}{2(2.3)(2.7)}$$

$$\cos\beta = \frac{5.29 + 7.29 - 2.25}{12.42}$$

$$\cos\beta = \frac{10.33}{12.42}$$

$$\cos\beta = 0.831$$

$\beta = \cos^{-1} 0.831$

$\beta = 33°43'$

$$\cos\gamma = \frac{a^2 + b^2 - c^2}{2ab}$$

$$\cos\gamma = \frac{(2.3)^2 + (1.5)^2 - (2.7)^2}{2(7)(9)}$$

$$\cos\gamma = \frac{5.29 + 2.25 - 7.29}{6.9}$$

$$\cos\gamma = \frac{0.25}{6.9}$$

$\cos\gamma = 0.036$

$\gamma = \cos^{-1} 0.036$

$\gamma = 87°55'$.

Example **In right triangle ABC, $\gamma = 90°$, a = 5, c = 13. Find the value of angle α.**

Solution: $\sin\alpha = \frac{a}{c}$

$$\sin\alpha = \frac{5}{13} = 0.3846$$

$\alpha = \sin^{-1}(0.3846)$

$\alpha = 22°37'$.

Example: **In the right triangle ABC, $\gamma = 90°, \alpha = 35°, a = 5$, find angle β.**

solution : $\alpha + \beta + \gamma = 180°$ $35° + \beta + 90° = 180°$

$\beta = 180° - 125°$

$\beta = 54°$.

Example: In the right triangle ABC, $\gamma = 90°$, , $\alpha = 35°$, $b = 6$, find side a

solution : $\tan \alpha = \frac{a}{b}$

$a = \tan 35° \times 6$

$a = 4.2$.

Example: In the right triangle ABC, $\alpha = 30°$, $c = 10$, $\gamma = 135°$ find a.

Solution: Using the law of sines:

$$\frac{a}{\sin \alpha} = \frac{c}{\sin \gamma}$$

$$\frac{a}{\sin 30} = \frac{10}{\sin 135°}$$

$$a = \frac{10 \times \sin 30°}{\sin 135°}$$

$a = 7.07$.

Example: In any triangle ABC, if $a = 20$, $c = 32$, $\gamma = 70°$, find α.

Solution: Using the law of sines:

$$\frac{a}{\sin \alpha} = \frac{c}{\sin \gamma}$$

$$\frac{20}{\sin \alpha} = \frac{32}{\sin 70°}$$

$$\sin \alpha = \frac{20 \times \sin 70°}{32}$$

$\sin \alpha = 0.587$

$\alpha = \sin^{-1} 0.587$

$\alpha = 35°57'$.

Example: **In any triangle ABC, if , $a = 9$, $b = 5$, $\gamma = 32°$, find c.**

Solution: Using the law of cosines.

$c^2 = a^2 + b^2 - 2ab\cos\gamma$

$c^2 = 9^2 + 5^2 - 2(9)(5)\cos 32°$

$c^2 = 29.67$. By taking square root:

$c = 5.45$.

Example: The sides of a triangle are 16, 20 and 33 m. Find its greatest angle.

Solution: Let $a = 16,\ b = 20,\ c = 33,\ \gamma = ?$

$$\cos\gamma = \frac{a^2 + c^2 - b^2}{2ac}$$

$$\cos\gamma = \frac{(16)^2 + (20)^2 - (33)^2}{2(16)(20)}$$

$$\cos\gamma = \frac{-433}{640}$$

$\cos\gamma = -0.6765$

$\gamma = \cos^{-1}(-0.6765)$

$\gamma = 132°34'32''$.

Law of Cotangent:

In trigonometry, the **lawss of cotangent** is a relationship among the lengths of the sides of a triangle and the cotangents of the halves of the three angles. Just as three quantities whose equality is expressed by the law of sines are equal to the diameter of the circumscribed circle of the triangle (or to its reciprocal, depending on how the law is expressed), the law of cotangents also relates the radius of the inscribed circle of a triangle (the inradius) to its sides and angles.

Statement: Using the usual notations for a triangle , where a,b,c are the lengths of the three sides, *A*, *B*, *C* are the vertices opposite those three respective sides, α,β,γ are the corresponding angles at those vertices, *s* is the semi-perimeter, that is, $s=\frac{a+b+c}{2}$, and r is the radius of the inscribed circle. The law of cotangents states that cot and furthermore that the inradius is given by:

$$\frac{\cot\frac{\alpha}{2}}{s-a}=\frac{\cot\frac{\beta}{2}}{s-b}=\frac{\cot\frac{\gamma}{2}}{s-c}.$$

Mollweide's formula

In trigonometry, Mollweide's formula, sometimes referred to in older texts as Mollweide's equations, named after Karl Mollweide, is a set of two relationships between sides and angles of a triangle. We can use this to check the solutions of triangles.

Let a, b, and c be the lengths of the three sides of a triangle. Let α, β, and γ be the measures of the angles opposite these three sides, respectively. Mollweide's formula states that:

$$\frac{a+b}{c}=\frac{\cos\frac{\alpha-\beta}{2}}{\sin\frac{\gamma}{2}}$$

$$\frac{a-b}{c}=\frac{\sin\frac{\alpha-\beta}{2}}{\cos\frac{\gamma}{2}}.$$

SUMMARY

In this chapter, we learned about the solution of triangles. There are some rules to solve an oblique triangle. The methods that we use to solve oblique triangles are the law of sines, law of cosines, law of tangents and half angle formulas. In terms of geometry, we learned that if two triangles are similar, the corresponding angles are congruent and the corresponding sides are in proportion. If the pairs of the corresponding angles and sides are congruent, then the triangles must be similar. The following pairs of congruent corresponding angles and proportional corresponding sides are sufficient to prove triangles similar:

1. Two angles (AA)
2. Three sides (SSS)
3. Two sides and the included angles (SAS)
4. Hypotenuse and one leg of a right triangle (HL)

After completing this chapter, a student will be able to:

- Understand types of triangles
- Obtain the solution of triangles using different methods
- Apply the law of sines
- Apply the law of cosines
- Apply the law of tangents
- Apply half angle formulas.

Key terms and formulas

Oblique Triangles

A triangle which is not right, is known as an **oblique triangle.**

Law of sines: If we know the measures of two angles and the included side of a triangle (ASA), or if we know the measures of two angles and an opposite angle of a triangle (AAS), the size and shape of the triangle is determined. Therefore, we should be able to find the measures of the remaining sides.

In any triangle. ABC

$$\frac{a}{\sin\alpha} = \frac{b}{\sin\beta} = \frac{c}{\sin\gamma}.$$

Law of cosines: When we know the measures of two sides and the included angle of a triangle (SAS), the size and shape of the triangle are determined. Therefore, we should be able to find the measure of the third side of the triangle.

In any triangle

$$a^2 = b^2 + c^2 - 2bc\cos\alpha$$

$$b^2 = c^2 + a^2 - 2ca\cos\beta$$

$c^2 = a^2 + b^2 - 2ab\cos\gamma$

$$\cos\alpha = \frac{b^2 + c^2 - a^2}{2bc}$$

$$\cos\beta = \frac{c^2 + a^2 - b^2}{2ca}$$

$$\cos\gamma = \frac{a^2 + b^2 - c^2}{2ab}.$$

Note: If Δ ABC is right, then the law of cosines reduces to Pythagoras Theorem i.e.

If $\alpha = 90°$ then $b^2 + c^2 = a^2$

Or If $\beta = 90°$ then $c^2 + a^2 = b^2$

Or If $\gamma = 90°$ then $a^2 + b^2 = c^2$

The law of tangents

$$\frac{a-b}{a+b} = \frac{\tan\left(\frac{\alpha-\beta}{2}\right)}{\tan\left(\frac{\alpha+\beta}{2}\right)}$$

$$\frac{b-c}{b+c} = \frac{\tan\left(\frac{\beta+\gamma}{2}\right)}{\tan\left(\frac{\beta-\gamma}{2}\right)}$$

$$\frac{c-a}{c+a} = \frac{\tan\left(\frac{\gamma-\alpha}{2}\right)}{\tan\left(\frac{\gamma+\alpha}{2}\right)}.$$

Half angle formulas

a) The sine of the half angle in terms of the sides

$$\sin\frac{\alpha}{2} = \sqrt{\frac{(s-b)(s-c)}{bc}}$$

$$\sin\frac{\beta}{2} = \sqrt{\frac{(s-c)(s-a)}{ca}}$$ where $2s = a + b + c$

$$\sin\frac{\gamma}{2}=\sqrt{\frac{(s-a)(s-b)}{ab}}.$$

b) The Cosine of the half angle in terms of the sides

In any triangle ABC

$$\cos\frac{\alpha}{2}=\sqrt{\frac{s(s-a)}{bc}}$$

$$\cos\frac{\beta}{2}=\sqrt{\frac{s(s-b)}{ac}}$$

$$\cos\frac{\gamma}{2}=\sqrt{\frac{s(s-c)}{ab}}$$

c) The tangent of the half angle in terms of the sides:

In any triangle ABC

$$\tan\frac{\alpha}{2}=\sqrt{\frac{(s-b)(s-a)}{s(s-a)}}$$

$$\tan\frac{\beta}{2}=\sqrt{\frac{(s-c)(s-a)}{s(s-b)}}$$

$$\tan\frac{\gamma}{2}=\sqrt{\frac{(s-a)(s-b)}{s(s-c)}}.$$

CHAPTER 7

AREA OF A TRIANGLE

INTRODUCTION

In the previous chapter, we discussed the solution of triangles. Now we will discuss the area of a triangle using different methods

AREA OF A TRIANGLE

The area of any triangle equals one-half of the product of its base and altitude. In general, if enough information about a triangle is known for its solution, then its area can be found.

Case I: Area of triangle in terms of the measure of two sides and their included angle.

The area of the triangle is equal to one-half of the product of the two sides times the sine of the included angle.

Statement: With usual notation, prove that:

Area of triangle ABC $= \frac{1}{2}bc\sin\alpha \quad = \frac{1}{2}ca\sin\beta \quad = \frac{1}{2}ab\sin\gamma$

Proof: Consider three different kinds of triangles ABC with m

i) acute ii) obtuse and iii) right

From A, draw

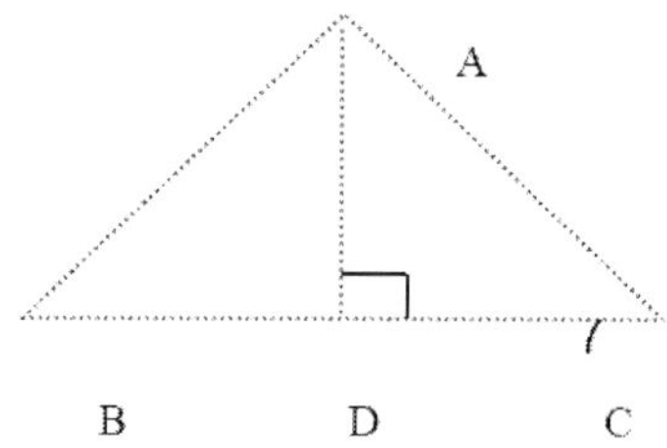

Figure i.

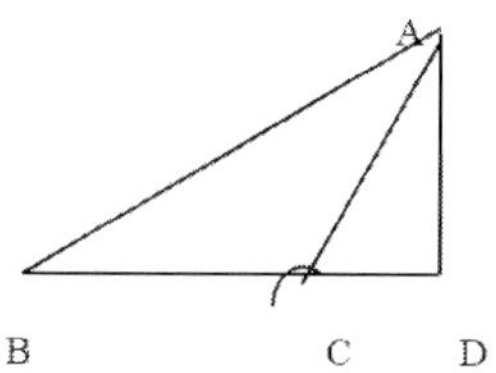

Figure ii.

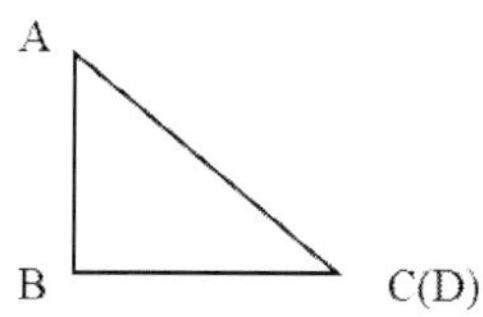

Fig iii.

Figure iii.s

In fig (i) $= \sin \gamma$.

In fig (ii) $\frac{AD}{AC} = \sin(180° - \gamma) = \sin \gamma$.

In fig. (iii), ss $\frac{AD}{AC} = 1 = \sin 90° = \sin \gamma$.

In all the three cases, we have:

AD = AC $\sin \gamma$ = b $\sin \gamma$.

Let Δ denote the area of triangle ABC.

By elementary geometry, we know that:

$\Delta = \frac{1}{2}$ (Base) (Altitude)

$\therefore \quad \Delta = \frac{1}{2} BC, AD$

$\therefore \quad \Delta = \frac{1}{2} ab \sin\gamma \cdot$

Similarly, we can prove that:

$$\Delta = \frac{1}{2} \sin\alpha = \frac{1}{2} ca \sin\beta$$

Case II. Area of triangle in terms of the measures of one side and two angles.

The third angle is found using the fact that . The area of the triangle equals a side squared times the product of the sines of the angles, including the side divided by twice the sine of the angle opposite the given side.

Statement: In a triangle with the usual notation, prove that:

$$\text{Area of triangle} = \frac{a^2 \sin\beta \sin\gamma}{2\sin\alpha} = \frac{b^2 \sin\gamma \sin\alpha}{2\sin\beta} = \frac{c^2 \sin\alpha \sin\beta}{2\sin\gamma}.$$

Proof: Using the law of sines, we know that:

$$\frac{\alpha}{\sin\alpha} = \frac{b}{\sin\beta} = \frac{c}{\sin\gamma}$$

$$\Rightarrow \quad a = c \cdot \frac{\sin\alpha}{\sin\gamma} \quad \text{and} \quad b = c \cdot \frac{\sin\beta}{\sin\gamma}.$$

We know that the area of triangle ABC is

$$\Delta = \frac{1}{2} ab \sin\gamma$$

$$\Rightarrow \Delta = \frac{1}{2}\left(c\frac{\sin\alpha}{\sin\beta}\right)\left(\frac{c \sin\beta}{\sin\gamma}\right) \sin\gamma$$

$$\Delta = \frac{c^2 \sin\alpha \sin\beta}{2\sin\gamma}.$$

In a similar way, we can prove that:

$\Delta ABC = \dfrac{a^2 \sin\beta \sin\gamma}{2\sin\alpha}$ and $\Delta ABC = \dfrac{b^2 \sin\gamma \sin\alpha}{2\sin\beta}$.

Case III: Area of triangle in terms of the measures of its sides.

The area of a triangle can be determined using the formula given below.

Statement: In a triangle ABC, with the usual notation, prove that:

Area of triangle $= \sqrt{s(s-a)(s-b)(s-c)}$.

Proof: We know that the area of a triangle ABC is

$$\Delta = \frac{1}{2}\text{bc}\sin\alpha$$

$$= \frac{1}{2}\text{bc } 2\sin\frac{\alpha}{2}\cos\frac{\alpha}{2} \quad \left(\therefore \sin\alpha = 2\sin\frac{\alpha}{2}\cos\frac{\alpha}{2}\right)$$

$$= \text{bc}\sqrt{\frac{(s-b)(s-c)}{bc}}\sqrt{\frac{s(s-a)}{bc}} \text{ (by the half angle formulas)}$$

$$= \text{bc}\sqrt{\frac{s(s-a)(s-b)(s-c)}{bc}}$$

$$\therefore \quad \Delta = \sqrt{s(s-a)(s-b)(s-c)}$$

Which is also called Hero's formulas

Example: Find the area of triangle ABC, in which b = 21.6, c=30.2 and $\alpha = 52°40$.

Solution: We Know that:

$$\Delta ABC = \frac{1}{2}\text{bc}\sin\alpha = \frac{1}{2}(21.6)(30.2)\sin 52°40$$

$= \frac{1}{2}(21.6)(30.2)(0.7951)$

$\therefore \Delta ABC \quad = \quad 259.3$ sq. units.

Example: Find the area of triangle ABC, whereby:

$\alpha = 35°17, \gamma = 45°13, and\, b = 42.1$

Solution: $\because \quad \alpha + \beta + \gamma = 180°$

$\because \quad \beta = 180° - (\alpha + \beta) = 180° - (35°17 + 45°13) = 99\,99°30$

Also b = 42.1, $\alpha = 35°17, \gamma = 45°13, \beta = 99°30$.

We know that the area of triangle ABC is:

$$\Delta = \frac{1}{2}\frac{b^2 \sin\gamma \sin\alpha}{\sin\beta}$$

$$\Delta = \frac{1}{2}\frac{(42.1)^2 \sin 45°13 \sin 35°17}{\sin 99°30}$$

$$= \frac{1}{2}\frac{(42.1)^2 (0.7097)(0.5776)}{0.9863}$$

$\Delta = 368.3$ sq. units.

Example: Find the area of the triangle ABC whereby:

$a = 275.4, b = 303.7, c = 342.5$

$$2s = a + b + c$$

$275.4 + 303.7 + 342.5 = 921.6$

$s = 460.8$.

Now $s - a = 460.8 - 275.4 = 185.4$

$s - b = 460.8 - 303.7 = 157.1$

s-c = 460.8 -342.5 =118.3.

Now $\Delta = \sqrt{s(s-a)(s-b)(s-c)}$

$= \sqrt{460.8 \text{ x}185.4 \text{ x}157.1\text{x}118.3}$

$\Delta = 39847$ sq.units.

APPLICATION

Problem 1. The hypotenuse of a right triangle is 10**cm** and its height is twice that of its base. Find the area of the triangle.

Solution: Given the hypotenuse of 10cm

Let x be the base of the right triangle and $2x$ its height.

By Pythagoras Theorem:

$(\text{Hypotenuse})^2 = (\text{Base})^2 + (\text{Prep})^2$

$10^2 = x^2 + (2x)^2$

$100 = x^2 + 4x^2$

$100 = 5x^2$

$20 = x^2$

$x = 4.47\,\text{cm}$

Height $= 2x = 2(4.47) = 8.94\,\text{cm}$

Area of triangle = $\frac{1}{2} \times$ Base $\times$ height $= \frac{1}{2}(4.47)(8.94) = 19.98$ cm^2.

Problem 2.Three perpendicular sides are drawn from a point within an equilateral at 6, 7 and 8 cm. Find the area of the triangle.

Solution: Let ABC be the equilateral triangle of side 'a' and 'O' be any point within the triangle.

OP = 6, OR = 7, OQ = 8 is the perpendiculars from O. Join O with A, B,

C.

Area of Δ ABC = Area of Δ AOB + Area of Δ BOC + Area of Δ AOC

Area of Δ ABC = $\frac{1}{2}$ (AB) (OP) + $\frac{1}{2}$ (BC) (OR) + $\frac{1}{2}$ (AC) (OQ)

$=\frac{1}{2}(a)(6)+\frac{1}{2}(a)(7)+\frac{1}{2}(a)(8)$

$=\frac{1}{2}(a)(21)$

$=\frac{21a}{2}$

$= 10.5a\ cm^2$

Area of equilateral triangle $=\frac{\sqrt{3}}{4}a^2$ cm^2

$=> \frac{\sqrt{3}}{4}a^2 = 10.5a$

$=> a = \frac{10.5\times4}{\sqrt{3}} = 24.25\,\text{cm}$

Area of ABC $=\frac{\sqrt{3}}{4}a^2 = \frac{\sqrt{3}}{4}(24.25)^2 = 254.60$ cm^2.

Problem 3. The sides of a triangular pond are 242, 1212 and 1450 m. Find the total amount of antiseptic medicine needed for spraying when one gallon of the medicine is sufficient for 100 m^2 of water surface of the pond.

Solution: Here $a = 242\,\text{m},\ b = 1212\,\text{m}, c = 1450\,\text{m}$

$$s = \frac{a+b+c}{2} = \frac{242+1212+1450}{2} = \frac{2904}{2} = 1452$$

Area $= \sqrt{s(s-a)(s-b)(s-c)}$

$$= \sqrt{1452(1452-242)(1452-1212)(1452-1450)}$$

$= \sqrt{1452 \times 1210 \times 240 \times 2}$

$= 29040$ m^2.

For 100 m^2, the amount of antiseptic medicine needed = 1 gallon.

For 1 m^2 = $\frac{1}{100}$ gallon.

$29040 \text{ m2} = \frac{1}{100} \times 29040 = 290.40$ gallon.

Problem 4. The sides of a triangle are 21, 20 and 13 cm. Find the area of the resulting triangles when the given triangle is divided by the perpendicular upon the longest side from the opposite angular point.

Solution: Let ABC be the triangle:

$a = BC = 21 \text{cm}$

$b = AC = 13 \text{cm}$

$c = AB = 20 \text{cm}$

$$s = \frac{a+b+c}{2} = \frac{21+13+20}{2} = \frac{54}{2} = 27$$

Area of ABC =

=

= cm^2.

Let AD be perpendicular on the longest side BC.

Area of Δ ABC $= \frac{1}{2} \times BC \times AD$

$126 = \frac{1}{2} \times 21 \times AD$

$AD = \frac{126 \times 2}{21}$

$AD = 12 \, cm$

In Δ ABD: $(AB)^2 = (BD)^2 + (AD)^2$

$(20)^2 = (BD)^2 + (12)^2$

$400 = (BD)^2 + 144$

$(BD)^2 = 400 - 144$

$(BD)^2 = 256$

$BD = 16\text{cm}$

As in figure: $BC = BD + DC$

$21 = 16 + DC$

$DC = 21 - 16$

$DC = 5\text{cm}$

Area of $\Delta ABD = \frac{1}{2} \times BD \times AD = \frac{1}{2} \times 16 \times 12 = 96\text{m}^2$

Area of $\Delta ACD = \frac{1}{2} \times DC \times AD = \frac{1}{2} \times 5 \times 12 = 30\text{m}^2$

Problem 5. From a point within a triangle, it is found that the three sides subtend equal angles. From this point, three lines are drawn to meet the opposite edges. If these lines measure 5, 6 and 7 cm, find the area of the triangle.

Solution: Let 'O' be the point within ΔABC.

In $\Delta ABC \angle AOB = \angle AOC = \angle BOC = \theta$

$\theta + \theta + \theta = 360^o$

$3\theta = 360^o$

$\theta = \frac{360^o}{3} = 120^\circ$

Area of Δ_{ABC} = Area of ΔAOB + Area of ΔBOC + Area of ΔAOC

$= \frac{1}{2}(5)(7)\sin\theta + \frac{1}{2}(5)(6)\sin\theta + \frac{1}{2}(6)(7)\sin\theta$

$= \frac{35}{2}\sin\theta + \frac{30}{2}\sin\theta + \frac{48}{2}\sin\theta$

$= \frac{\sin\theta}{2}(35 + 30 + 42)$

$= \frac{\sin\theta}{2}(107)$

$= 53.5\sin\theta$

$= 53.5 \times \sin 120^\circ$

$= 46.33$ cm^2.

Problem 6. The sides of a triangular lawn are proportional to the numbers 5, 12 and 13. The cost of fencing the lawn at the rate of Rs.2 per meter is Rs.120. Find the lengths of the sides. Also find the cost of turfing the lawn at 25 paisa per square meter.

Solution: Let a,b,c be the sides of the triangular lawn.

As given $a \propto 5 \Rightarrow a = 5k, b \propto 12 \Rightarrow b = 12k, c \propto 13 \Rightarrow c = 13k$

Perimeter $= a + b + c = 5k + 12k + 13k = 30k$

Cost of fencing = Rate $\times$ Perimeter

$120 = 2 \times$ Perimeter

$\frac{120}{2}$ = Perimeter

Perimeter $= 60$

$30k = 60$

$k = \frac{60}{30} = 2$

$a = 5(2) = 10$

$b = 12(2) = 24$

$c = 13(2) = 26$

$s = \frac{a+b+c}{2} = \frac{10+24+26}{2} = \frac{60}{2} = 30$

Area of lawn $= \sqrt{s(s-a)(s-b)(s-c)}$

$= \sqrt{30(30-10)(30-24)(30-26)}$

$= \sqrt{30 \times 20 \times 6 \times 4} = 120\, sq.m$

Cost of turfing= Area $\times$ Rate $= 120 \times \frac{25}{100} = 30\, rupees$.

Problem 7. Find the area of the triangle whose sides have the ratio 9:40:41 and whose perimeter in 180 m.

Solution: Let a,b,c are the sides of triangle.

Given $a:b:c = 9:40:41$

$a \propto 9 \Rightarrow a = 9k, b \propto 40 \Rightarrow b = 40k, c \propto 41 \Rightarrow c = 41k$

Given perimeter $= 180$

$a + b + c = 180$

$9k + 40k + 41k = 180$

$90k = 180$

$k = \frac{180}{90} = 2$

$a = 9(2) = 18$

$b = 40(2) = 80$

$c = 41(2) = 82$

$s = \frac{a+b+c}{2} = \frac{180}{2} = 90$

Area of triangle $= \sqrt{s(s-a)(s-b)(s-c)}$

$= \sqrt{90(90-18)(90-80)(90-82)}$

$= \sqrt{90(72)(10)(8)}$

$= 720\,\text{cm}^2.$

Problem 8. The corresponding bases of two triangles, at an equal altitude, are 8 cm and

10 cm. The area of the smaller triangle is 108 cm^2. Find the area of the larger triangle and altitude of each.

Solution: Let a =base of smaller triangle $= 8\,\text{cm}$, $b =$ base of large triangle $= 10\,\text{cm}$

Area of smaller triangle $= 108$ cm^2.

And h = altitude of each triangle

Area of smaller triangle $= \frac{1}{2} \times a \times h$

$108 = \frac{1}{2} \times 8 \times h$

$h = \frac{108 \times 2}{8}$

$h = 27$.

Area of larger triangle $= \frac{1}{2} \times b \times h = \frac{1}{2} \times 10 \times 27 = 135$ cm^2.

CIRCLES CONNECTED WITH A TRIANGLE

In our previous classes, we learnt the methods of drawing the following three kinds of circles related to a triangle:

i) Circum–Circle

ii) In Circle

iii) Ex- Circle.

Circum-Circle:

The Circle passing through the three vertices of a triangle is called a Circum-Circle. Its center is called the circum-center, and is the point of the intersection of the bisectors of the sides of the triangle. Its radius is known as the circum-radius and is denoted by R.

Prove that: $R = \frac{a}{2\sin\alpha} = \frac{b}{2\sin\beta} = \frac{c}{2\sin\gamma}$, with the usual notation.

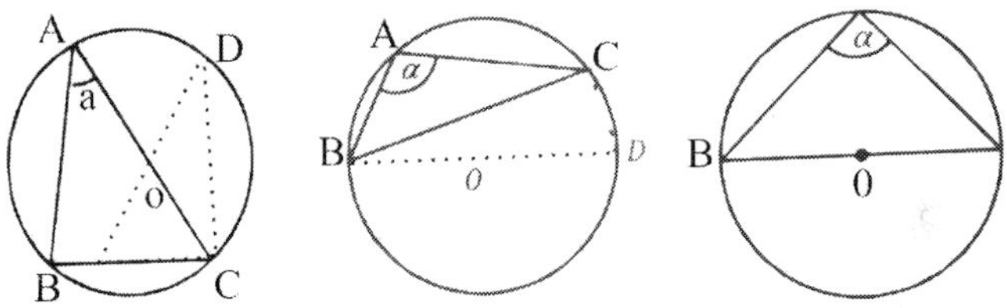

Figure 13

Proof: Consider three different kinds of triangles ABC with m$> A = a$

i) acute

ii) obtuse

iii) right.

Let O be the circum-center of ΔABC, join B to O and produced $\overline{BO}$ to meet the circle again at D. Join C to D. Thus, we have the measure of diameter

m $\overline{BD}$ = 2R and m $\overline{BC}$ = a.

I. In fig (i), $BDC = m\angle A = \alpha$ angle in tshe same segment in the right triangle BCD.

$$\frac{\overline{BC}}{BD} = \sin m\angle BDC = \sin\alpha .$$

II. In fig. (ii), m∠ BDC + m∠ A = 180° (sum of opposite angles of a)

$\Rightarrow$ $m\angle BDC + \alpha = 180°$ cyclic quadrilateral = 180°

$\Rightarrow$ $m\angle BDC = 180° - \alpha.$

In the right triangle BCD,

$$\frac{m\overline{BC}}{mBD} = \sin m_{\angle BCD = \sin(180° - \alpha)} = \sin\alpha .$$

III. In fig.(iii), $m\angle A = \alpha = 90°$

$$\frac{m\overline{BC}}{mBD} = 1 = \sin 90° = \sin\alpha.$$

In all the three figures, we have proved that

$$\frac{m\overline{BC}}{mBD} = \sin\alpha$$

$$\Rightarrow \quad \frac{\alpha}{2R} = \sin\alpha \Rightarrow 2R\sin\alpha = a$$

$$\therefore \quad R = \frac{a}{2\sin\alpha}$$

Similarly, we can prove that

$$R = \frac{b}{2\sin\beta} \text{ and } R = \frac{R.c}{2\sin\gamma}.$$

$$\text{Hence, } R = \frac{a}{2\sin\alpha} = \frac{b}{2\sin\beta} = \frac{c}{2\sin\gamma}.$$

a) Deduction of the law of sines:

We know that $R = \dfrac{\alpha}{2\sin\alpha} = \dfrac{b}{2\sin\beta} = \dfrac{c}{2\sin\gamma}$

$$= \frac{\alpha}{\sin\alpha} = \frac{\beta}{\sin\beta} = \frac{c}{\sin\gamma} = 2R$$

$\Rightarrow \quad \frac{\alpha}{\sin\alpha} = \frac{\beta}{\sin\beta} = \frac{c}{\sin\gamma},$ which is the law of sines.

b) Prove that: $R = \frac{abc}{4\Delta}$

Proof: We know that: $R = \frac{\alpha}{2\sin\alpha}$

$$\Rightarrow \quad R = \frac{a}{2\,.2\sin\frac{\alpha}{2}\cos\frac{\alpha}{2}} \qquad \left(\because \sin\alpha = 2\sin\frac{\alpha}{2}\cos\frac{\alpha}{2}\right)$$

$$= \frac{a}{4\sqrt{\frac{(s-b)(s-c)}{bc}}\sqrt{\frac{\sqrt{s(s-a)}}{bc}}}$$

$$\therefore R = \frac{abc}{4\Delta}$$

$$\because \Delta = \sqrt{s(s-a)(s-b)(s-c)}.$$

In-Circle

The circle drawn inside a triangle touching its three sides is called its inscribed circle or in- circle. Its center, known as the in center is the point of intersection of the bisectors of angles of triangle. Its radius is called in radius and is denoted by r.

Prove that: $r = \frac{\Delta}{s}$ with usual notations.

Proof: Let the internal bisector of angles of triangle ABC meet at *O*, the in-center

Draw $\overline{OD} \perp \overline{BC}$, $\overline{OE} \perp \overline{AC}$ and $\overline{OF} \perp \overline{AB}$

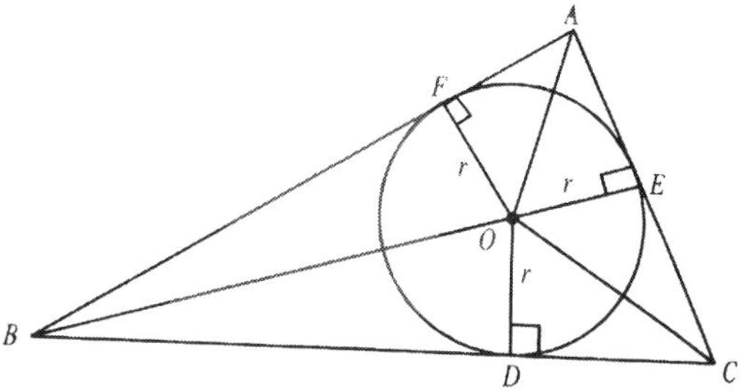

Figure 14

Let $m\ \overline{OD} = m\ \overline{OE} = m\ \overline{OF} = r$

From the figure Area ΔABC = Area ΔOBC + Area ΔOCA + Area ΔOAB

$$\Delta = \frac{1}{2} BC \times OD + \frac{1}{2} CA \times OE + \frac{1}{2} + AB \times OF$$

$$= \frac{1}{2}ar + \frac{1}{2}br + \frac{1}{2}cr$$

$$= \frac{1}{2}r(a+b+c)$$

$$\Delta = \frac{1}{2} r.\ 2s \qquad \because 2s = (a+b+c)$$

Escribed Circles

A Circle, which touches one side of the triangle externally and the other two produced sides, is called an escribed circle or ex-circle or e-circle. Obviously there could be only three such circles of a triangle, one opposite to each angle of triangle. The centers of these circles, which are called ex-center are the point of the internal bisector of one and external bisectors of one and the external bisectors of the other two angle triangle meet.

In ΔABC, centre of the ex-circles opposite to the vertex A is usually take and its radius is denoted by r_1. Similarly, centers of ex-circles opposite to the vertex B and C are taken as l_2 and l_3 and their radii are denoted by r_2 and r_3 respectively

a) With usual notation, prove that:

$r_1 = \frac{\Delta}{s-a}$, $r_2 = \frac{\Delta}{s-b}$, and $r_3 = \frac{\Delta}{s-c}$

Proof: Let l_1, draw $\overline{l_1 D} \perp, \overline{BC}$, $\overline{l_1 E} \perp \overline{AC}$ Produced

And $\overline{l_1 F} \perp \overline{AB}$ Produced

Join l_1 to A, B and C

Let m $\overline{l_1 D} = m\overline{l_1 E} = m\overline{l_1 F} = r_1$

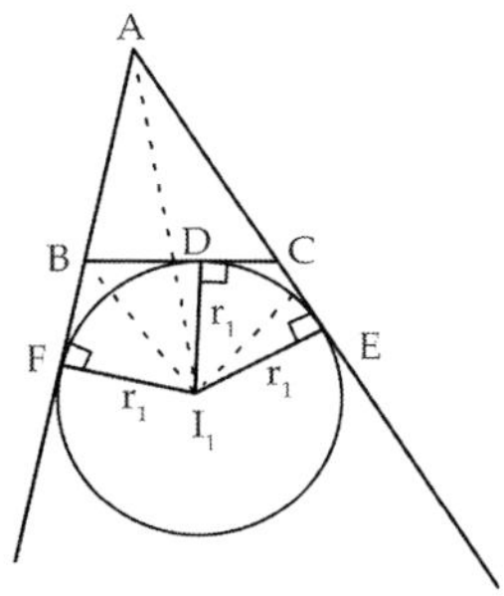

Figure 15

From the figure $\Delta ABC = \Delta l_1 AB + \Delta l_1 AC - \Delta l_1 BC$

$\Delta = \frac{1}{2} AB x l_1 F + \frac{1}{2} AC x l_1 E - \frac{1}{2} BC x l_1 D$

$= \frac{1}{2} c\, r_1 + \frac{1}{2} b r_1 - \frac{1}{2} a\, r_1$

$\Delta = r_1 (c+b+a)$

$= \frac{1}{2} r_1 2(s-a) \; (2s = a+b+c)$

$= (s-a) r_1$

Hence $r_1 = \frac{\Delta}{s-a}$

In a similar way, we can prove that:

$r_2 = \frac{\Delta}{s-b}$ and $r_3 = \frac{\Delta}{s-c}$

Example: Show that

$$r=(s-a)\tan\frac{\alpha}{2}=(s-b)\tan\frac{\beta}{2}=(s-c)\tan\frac{\gamma}{2}$$

Solution: To prove $r=(s-a)\tan\frac{\alpha}{2}$

We know that: $\tan\frac{\alpha}{2}=\sqrt{\frac{(s-b)(s-c)}{s(s-a)}}$

RHS = $(s-a)\tan\frac{\alpha}{2}=(s-a)\sqrt{\frac{(s-b)(s-c)}{s(s-a)}}$

$$=\sqrt{\frac{(s-a)(s-b)(s-c)}{s}}$$

$$=\sqrt{\frac{s(s-a)(s-b)(s-c)}{s^2}}=\frac{\Delta}{s}=r,$$

In a similar way, we can prove that:

$r=(s-b)=\tan\frac{\beta}{2}$ and $r=(s-c)\tan\frac{\gamma}{2}$.

Example: Show that $r_1 = 4R\sin\frac{\alpha}{2}\cos\frac{\beta}{2}\cos\frac{\gamma}{2}$.

Solution: RHS. $= 4R\sin\frac{\alpha}{2}\cos\frac{\beta}{2}\cos\frac{\gamma}{2}$.

$$=4.\frac{abc}{4\Delta}\sqrt{\frac{(s-b)(s-c)}{bc}}\sqrt{\frac{s(s-b)}{ca}}\sqrt{\frac{s(s-c)}{ab}}$$

$$=\frac{s(s-a)(s-c)}{\Delta}$$

$$= \frac{s(s-a)(s-b)(s-c)}{\Delta .(s-a)}$$

$$= \frac{\Delta^2}{\Delta(s-a)}$$

$$= \frac{\Delta}{s-a} = r_1 \quad = \text{LHS.}$$

ENGINEERING AND CIRCLES CONNECTED WITH TRIANGLES

We know that frames of all rectilinear shapes, with the exception of triangular ones, change their shapes when pressed from two corners. The fact that a ssstriangular frame does not change its shape when it is pressed from any vertices means that a triangle is the only rigid rectilinear figure. It is on this account that engineers make frequent use of triangles to strengthen material in all sorts of construction work. Besides triangular frames etc., circular rings can stand greater pressure when pressed from any two points. That is why wells are always made cylindrical, with a circular surface that can stand the surrounding pressure from water. Moreover, arches below bridges are constructed in the shape of arcs of circles so that they can bear the burden of traffic passing over the bridge.

a) We know that triangular frames change their rectilinear nature when they are pressed from the sides. Engineers have to fix circular rings touching the sides of the triangular frames. In order to construct these rings, the in-radii of the triangles need to be determined.

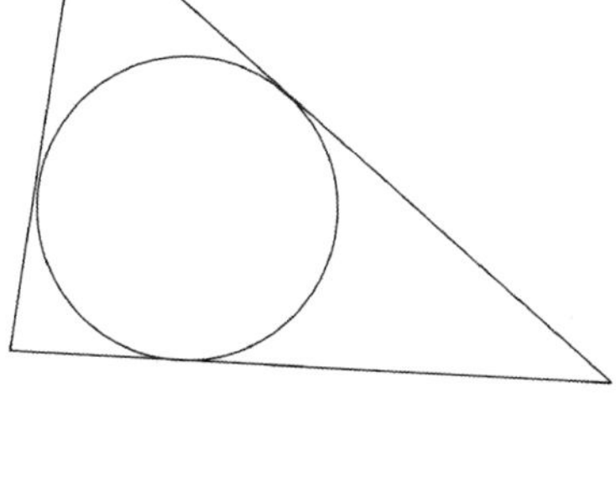

Figure 16

b) In order to protect the triangular discs from any kind of damage, engineers fit circular rings enclosing the discs. To obtain the proper size of the rings, engineers are bound to calculate the circum-radii of the triangles.

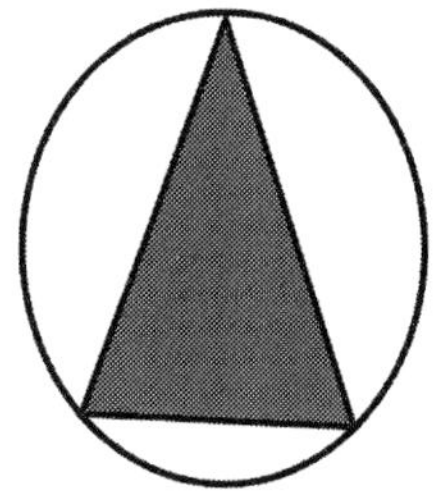

Figure 17

c) In certain triangular frames, engineers have to extend two sides of the frames. In order to strengthen these loose wings, engineer feel the necessity to fix circular rings touching the extended sides and the third side of the frames.

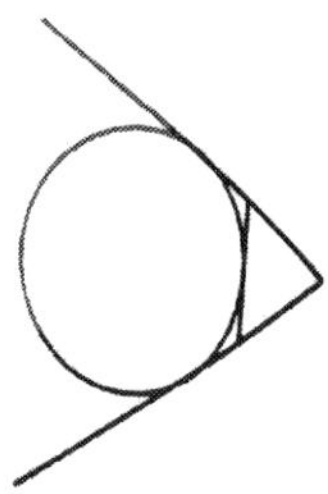

Figure 18

In order to make the appropriate rings, engineers have to find ex-radii of the triangles.

The above discussion shows that an engineer must know how to calculate the radii of an in-circle, circum-circle and an ex-circle of a triangle.

Problem 1: Prove that $r = 4R \sin\frac{\alpha}{2}\sin\frac{\gamma}{2}$.

RHS $= 4R \sin\frac{\alpha}{2}\sin\frac{\gamma}{2}$

$= 4R\sqrt{\frac{(s-b)(s-c)}{bc}}\sqrt{\frac{(s-a)(s-c)}{ac}}\sqrt{\frac{(s-a)(s-b)}{sb}}$

$$= 4R\sqrt{\frac{(s-b)(s-c)(s-a)(s-c)(s-a)(s-b)}{(bc)(ac)(ab)}}$$

$$= 4R\ \frac{(s-a)(s-b)(s-c)}{a^2b^2c^2}$$

$$= 4\left(\frac{abc}{4\Delta}\right)\ \frac{(s-a)(s-b)(s-c)}{a^2b^2c^2}$$

$$= \frac{(s-a)(s-b)(s-c)}{\Delta}\quad \frac{s(s-a)(s-b)(s-c)}{\Delta}$$

$$= \frac{\Delta^2}{s\Delta}\quad \because \Delta = \sqrt{s(s-a)(s-b)(s-c)}$$

$$= \frac{\Delta}{s} = r = \text{LHS.}$$

Problem 2: Prove that $s = 4\,R\cos\frac{\alpha}{2}\ \cos\frac{\beta}{2}\cos\frac{\gamma}{2}$.

$$\text{RHS} = 4\,R\cos\frac{\alpha}{2}\ \cos\frac{\beta}{2}\cos\frac{\gamma}{2}$$

$$= 4R\sqrt{\frac{s(s-a)}{bc}}\ \sqrt{\frac{s(s-b)}{ac}}\ \sqrt{\frac{s(s-c)}{ab}}$$

$$= 4R\sqrt{\frac{s^2.\, s(s-a)(s-b)(s-c)}{(bc)(ac)(ab)}}\quad = 4R\sqrt{\frac{s^2\Delta^2}{a^2b^2c^2}}$$

$$= 4R\ \frac{s\Delta}{abc} = 4\left(\frac{abc}{4\Delta}\right)\text{S}\ \frac{\Delta}{abc}$$

$$\because R = \frac{abc}{4\Delta}$$

$$= S$$

Problem 3: Prove that $r = a\ \sin\frac{\beta}{2}\sin\frac{\gamma}{2}\sec\frac{\alpha}{2}$.

We take $a\sin\frac{\beta}{2}\sin\frac{\gamma}{2}\ \sec\frac{\alpha}{2}$

$=$

$$a\sin\frac{\beta}{2}\sin\frac{\gamma}{2}\frac{1}{cos\frac{\alpha}{2}}$$

$$= a\ \sqrt{\frac{(s-b)(s-c)}{ac}}\sqrt{\frac{(s-a)(s-b)}{ab}}\ \text{-}\ \frac{1}{\sqrt{\frac{s(s-a)}{bc}}}$$

$$= a\ \sqrt{\frac{(s-b)(s-c)}{ac}}\sqrt{\frac{(s-a)(s-b)}{ab}}\sqrt{\frac{bc}{s(s-a)}}$$

$$= a\ \sqrt{\frac{(s-a)(s-c)(s-a)(s-b)(bc)}{(ac)(ab)s(s-a)}}$$

$$= a\ \sqrt{\frac{(s-a)(s-b)(s-c)}{a^2b}}$$

$$= a\ \sqrt{\frac{s(s-a)(s-b)(s-c)}{a^2s^2}}$$

$$=a\ \sqrt{\frac{s(s-a)(s-b)(s-c)}{as}}$$

$$= \frac{\Delta}{s} = r$$

$$\Rightarrow\ a\sin\frac{\beta}{2}\ sin\frac{\gamma}{2}\ sec\frac{\alpha}{2} = r\ .$$

Similarly, we can prove the other parts.

$$bsin\frac{\gamma}{2}\sin\frac{\alpha}{2}sec\frac{\beta}{2} = r..........(ii)$$

$$c\sin\frac{a}{2}\ \sin\ \frac{\beta}{2}sec\frac{\gamma}{2} = r............(iii)$$

From (i), (ii) and (iii):

$$r = a\sin\frac{\beta}{2}sin\frac{\gamma}{2}sec\frac{\alpha}{2} = b\sin\frac{\gamma}{2}\sin\frac{\alpha}{2}sec\frac{\beta}{2} = c\sin\frac{\alpha}{2}\sin\frac{\beta}{2}\sec\frac{\gamma}{2}.$$

Problem 4: Prove that $r_1 = 4\ \mathrm{R} \sin \frac{\alpha}{2} \cos\frac{\beta}{2} \cos\frac{\gamma}{2}$

RHS = $4\ \mathrm{R} \sin \frac{\alpha}{2} \cos\frac{\beta}{2} \cos\frac{\gamma}{2}$

$$= 4\mathrm{R}\ \sqrt{\frac{(s-b)(s-c)}{bc}}\sqrt{\frac{s(s-b)}{ac}}\sqrt{\frac{s(s-c)}{ab}}$$

$$= 4\mathrm{R}\ \sqrt{\frac{(s-b)(s-b)s(s-b)s(s-c)}{(bc)(ac)(ab)}}$$

$$= 4\mathrm{R}\ \sqrt{\frac{s^2(s-b)^2(s-c)^2}{a^2b^2c^2}}$$

$$= 4\mathrm{R}\ \ \frac{s(s-b)(s-c)}{abc}$$

$$=4\,\frac{abc}{4\Delta}\,\frac{abc\,s(s-b)(s-c)(s-a)}{abc}$$

$\because\ R = \frac{abc}{4\Delta}$

$$= \frac{s(s-a)(s-b)(s-c)}{\Delta(s-a)}$$

$$= \frac{\Delta^2}{\Delta(s-a)}$$

$$= \frac{\Delta}{(s-a)}$$

$= r_1$ = RHS.

Problem 5: Prove that $r_1 = s\ tan\frac{a}{2}$

RHS = s tan $\frac{a}{2}$

$$= s\sqrt{\frac{(s-b)(s-c)}{s(s-a)}} = s\sqrt{\frac{(s-b)(s-c)s(s-a)}{s(s-a)\ s(s-a)}}$$

$$= \sqrt{\frac{s(s-a)(s-b)(s-c)}{s^2(s-a)^2}}$$

$$= \text{s} \sqrt{\frac{\Delta^2}{s^2(s-a)^2}}$$

$$= \text{s} \frac{\Delta}{s(s-a)}$$

$$= \frac{\Delta}{(s-a)}$$

$= r_1 =$ LHS.

Problem 6: Prove that $r_1 r_2 + r_2 r_3 + r_3 r_1 = s^2$

LHS = $r_1 r_2 + r_2 r_3 + r_3 r_1$

$$= \left(\frac{\Delta}{s-a}\right)\left(\frac{\Delta}{s-b}\right) + \left(\frac{\Delta}{s-b}\right)\left(\frac{\Delta}{s-c}\right) + \left(\frac{\Delta}{s-c}\right)\left(\frac{\Delta}{s-a}\right)$$

$$= \frac{\Delta^2}{(s-a)(s-b)} + \frac{\Delta^2}{(s-b)(s-c)} + \frac{\Delta^2}{(s-c)(s-a)}$$

$$= \Delta^2 \left(\frac{1}{(s-a)(s-b)} + \frac{1}{(s-b)(s-c)} + \frac{1}{(s-c)(s-a}\right)$$

$$= \Delta^2 \left(\frac{s-c+s-a+s-b}{(s-a)(s-b)(s-c)}\right) = \Delta^2 \left(\frac{3s-(a+b+c)}{(s-a)(s-b)(s-c)}\right)$$

$$= \Delta^2 \left(\frac{3s-2s}{(s-a)(s-b)(s-c}\right) \qquad \because s = \frac{a+b+c}{2}$$

$$= \Delta^2 \left(\frac{s}{(s-a)(s-b)(s-c} \frac{s}{s}\right)$$

$$= \Delta^2 \frac{s^2}{(s-a)(s-b)(s-c)} = \Delta^2 \left(\frac{s^2}{\Delta^2}\right) = s^2 = \text{RHS.}$$

Problem 7: Prove that $r r_1 r_2 r_3 = \Delta^2$

LHS = $r r_1 r_2 r_3$

$$=\left(\frac{\Delta}{s}\right)\left(\frac{\Delta}{s-a}\right)\left(\frac{\Delta}{s-b}\right)\left(\frac{\Delta}{s-c}\right)$$

$$=\frac{\Delta^4}{s(s-a)(s-b)(s-c)}$$

$$=\frac{\Delta^4}{\Delta^2}$$

$= \Delta^2 =$ RHS.

Problem 8: Prove that $r_1 + r_2 + r_3 - r = 4R$

LHS = $r_1 + r_2 + r_3 - r$

$$=\frac{\Delta}{s-a}+\frac{\Delta}{s-b}+\frac{\Delta}{s-c}-\frac{\Delta}{s}$$

$$=\Delta\left(\frac{1}{s-a}+\frac{1}{s-b}+\frac{1}{s-c}-\frac{1}{s}\right)$$

$$=\Delta\left(\frac{(s-b)+(s-a)}{(s-a)(s-b)}++\frac{s-(s-c)}{s(s-c)}\right)$$

$$=\Delta\left(\frac{2s-b-a}{(s-a)(s-b)}+\frac{s-s+a}{s(s-c)}\right)$$

$$=\Delta\left(\frac{a+b+c-b-a}{(s-a)(s-b)}+\frac{c}{s(s-c)}\right) \qquad \because 2s=a+b+c$$

$$=\text{c}\,\Delta\left(\frac{C}{(s-a)(s-b)}+\frac{C}{s(s-c)}\right)$$

$$=\text{c}\,\Delta\left(\frac{1}{(s-a)(s-b)}+\frac{1}{s(s-c)}\right)$$

$$=\text{c}\,\Delta\left(\frac{s(s-c)-(s-a)(s-b)}{s(s-a)(s-b)s-c)}\right)$$

$$=c\Delta\frac{s^2-sc+s^2-as-bs+ab}{\Delta^2}$$

$$=c\left(\frac{2s^2-s(a+b+c)+ab}{\Delta}\right)$$

$$=c\left(\frac{2s^2-s(2s)+ab}{\Delta}\right)$$

$$=c\left(\frac{2s^2-2s^2+ab}{\Delta}\right)$$

$$=\frac{abc}{\Delta}$$

$$=4.\frac{abc}{4\Delta}$$

$= 4R =$ RHS.

Problem 9: Prove that $r_1r_2r_3 = rs^2$.

$$\text{LHS} = \left(\frac{\Delta}{s-a}\right)\left(\frac{\Delta}{s-b}\right)\left(\frac{\Delta}{s-c}\right)$$

$$=\frac{\Delta^3}{(s-a)(s-b)(s-c)}$$

$$=\frac{s\Delta^3}{s(s-a)(s-b)(s-c)}$$

$$=\frac{s\Delta^3}{\Delta^2}$$

$$= s\Delta$$

$$=s^2\frac{\Delta}{s}$$

$= s^2r =$ RHS.

Problem 10: Find R, r, r_1, r_2 and r_3, if the measures of the sides of the triangle ABC are:

(i) a=13, b = 14, c = 15

$$s=\frac{a+b+c}{2}=\frac{13+14+15}{2}=21$$

$$s-a=21-13=8$$

$$s-b=21-14=7$$

$$s-c=21-15=6$$

So $\Delta=\sqrt{s(s-a)(s-b)(s-c)}$

$$=\sqrt{21(8)(7)(6)}$$

$$=\sqrt{7056}=84.$$

Now

$$R=\frac{abc}{4\Delta}=\frac{(13)(14)(15)}{4(84)}=8.125$$

$$r=\frac{\Delta}{s}=\frac{84}{21}=4$$

$$r_1=\frac{\Delta}{s-a}=\frac{84}{8}=10.5$$

$$r_2=\frac{\Delta}{s-b}=\frac{84}{7}=12$$

$$r_3=\frac{\Delta}{s-c}=\frac{84}{6}=12$$

SUMMARY

After completing this chapter, a student will be able to:

Understand the concept of area.

Know different methods for calculating the area.

Apply all formulas to different problems.

Understand the relationship between a circle and a triangle.

Understand the basic concept of a Circum circle.

Understand basic concept of an In circle.

Understand basic concept of an Escribed circle.

Review of common terms and basic formulas

Case I Area of a triangle in terms of the measure of two sides and their angle:

Area of triangle ABC $= \frac{1}{2} bc \sin\alpha = \frac{1}{2} ca \sin\beta = \frac{1}{2} ab \sin\gamma.$

Case II. Area of triangle in terms of the measures of one of the sides:

Area of triangle = $\frac{a^2 \sin\beta \sin\gamma}{2\sin\alpha} = \frac{b^2 \sin\gamma \sin\alpha}{2\sin\beta} = \frac{c^2 \sin\alpha \sin\beta}{2\sin\gamma}.$

Case III: Area of triangle in terms of the measures of its sides:

Area of triangle = $\sqrt{s(s-a)(s-b)(s-c)}$.

This is also called Hero's formulas, where $s = \frac{a+b+c}{2}$

Area of an Equilateral triangle $= \frac{\sqrt{3}}{4} a^2$

Cost of turfing = Area × Rate.

Circum-Circle:

The circle passing through the three vertices of a triangle is called a Circum-Circle. Its center is known as the circum-center; the point of the intersection of the bisectors of the sides of the triangle. Its radius is called the circum-radius, denoted by R.

$$R = \frac{a}{2\sin\alpha} = \frac{b}{2\sin\beta} = \frac{c}{2\sin\gamma}$$

$$R = \frac{abc}{4\Delta}$$

In-Circle

The circle drawn inside a triangle touching its three sides is called its inscribed circle or in- circle. Its center, known as the in-center, is the point of intersection of the bisectors of angles of the triangle. Its radius is called the in-radius, and is denoted by r.

$$r = \frac{\Delta}{s}$$

Escribed Circles

A circle which touches one side of the triangle externally and the other two produced sides, is called an escribed circle or ex-circle or e-circle. There can only be three such circles for a triangle, one opposite to each angle of the triangle. The centers of these circles, the ex-center, represent the point of the internal bisector of one and external bisectors of one and the external bisectors of the other two angle triangle meet.

In $\Delta ABC,$ the center of the ex-circles opposite to vertex A is usually taken, and its radius denoted by r_1. Similarly, centers of ex-circles opposite to the vertex B and C are taken as l_2 and l_3 and their radii are denoted by r_2 and r_3 respectively

$$r_1 = \frac{\Delta}{s-a}, \; r_2 = \frac{\Delta}{s-b}, \text{ and } r_3 = \frac{\Delta}{s-c}.$$

CHAPTER 8

TRIGONOMETRIC FUNCTIONS AND THEIR GRAPHS

DOMAINS AND RANGES OF SINE AND COSINE FUNCTIONS

We have already defined the trigonometric functions of $\sin\theta, \cos\theta, \tan\theta, \csc\theta, \sec\theta$ and $\cot\theta$.

We know that if $P(x, y)$ is any point on the unit circle with the center at the origin O, such that $\angle XOP = \theta$ is in standard position, then $\cos\theta = x$ and $\sin\theta = y$:

$\Rightarrow$ for any real number θ there is one and only one value of each x and y i.e., for each $\cos\theta$ and $\sin\theta$.

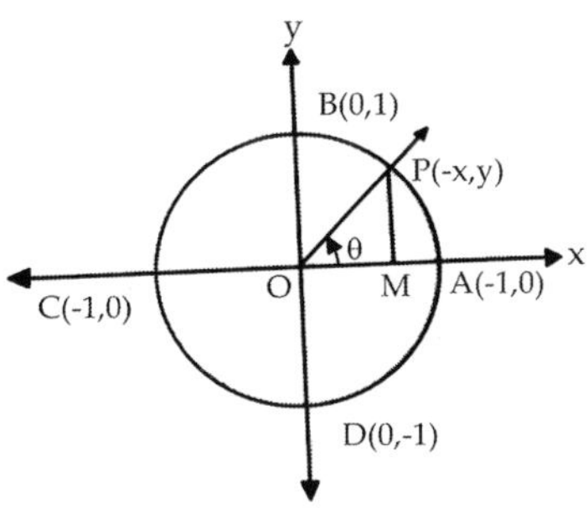

Figure 1.

Hence, $\sin\theta$ and $\cos\theta$ are the functions of θ and their domain is $\mathbb{R}$, the set of all real numbers.

Since $P(x, y)$ is a point on the unit circle with its center at the origin O,

$\therefore\ -1 \le x \le 1$ and $-1 \le y \le 1$

$\Rightarrow\ -1 \le \cos\theta \le 1$ and $-1 \le \sin\theta \le 1$.

Thus, the range of both the sine and cosine functions is $[-1, 1]$.

DOMAINS AND RANGES OF TANGENT AND COTANGENT FUNCTIONS

i) From the above figure:

$\tan\theta = \frac{y}{x}, \qquad x \ne 0$.

$\Rightarrow$ Terminal side $\overrightarrow{OP}$ should not coincide with OY or OY' (i.e.the Y – axis).

$\Rightarrow\ \theta \ne \pm\frac{\pi}{2}, \pm\frac{3\pi}{2}, \pm\frac{5\pi}{2}, \ldots$

$\Rightarrow\ \theta \ne (2n+1).\frac{\pi}{2}$, where $n \in Z$.

$\therefore$ The domain of the tangent function $= \mathbb{R} - \{x \mid x = (2n+1).\frac{\pi}{2}, \quad n \in Z\}$

and the range of the tangent function $= \mathbb{R} =$ set of real numbers.

ii) From the figure above:

$\cot\theta = \frac{x}{y}, \qquad y \ne 0$

$\Rightarrow$ Terminal side $\overrightarrow{OP}$ should not coincide with OX or OX' (i.e. the X – axis).

$\Rightarrow \theta \neq 0, \pm\pi, \pm 2\pi, \ldots$

$\Rightarrow \theta \neq n\pi$, where $n \in Z$.

$\therefore$ The domain of the cotangent function $= \mathbb{R} - \{x | x = n\pi,\ n \in Z\}$

and the range of the cotangent function $= \mathbb{R} =$ set of real numbers.

DOMAIN AND RANGE OF THE SECANT FUNCTION

From the above figure:

$\sec\theta = \frac{1}{x}$, $x \neq 0$

$\Rightarrow$ Terminal side $\overrightarrow{OP}$ should not coincide with OY or OY' (i.e.the Y – axis).

$\Rightarrow \theta \neq \pm\frac{\pi}{2}, \pm\frac{3\pi}{2}, \pm\frac{5\pi}{2}, \ldots$

$\Rightarrow \theta \neq (2n+1).\frac{\pi}{2}$, where $n \in Z$.

$\therefore$ The domain of the secant function $= \mathbb{R} - \{x | x = n\pi,\ n \in Z\}$.

As $\sec\theta$ attains all real values except those between -1 and ,

$\therefore$ The range of the secant function $= \mathbb{R} - \{x | -1 < x < 1\}$.

Domain and range of the cosecant function

From the above figure:

$\csc\theta = \frac{1}{y}$, $y \neq 0$

$\Rightarrow$ Terminal side $\overrightarrow{OP}$ should not coincide with OX or OX' (i.e.the X – axis).

$\Rightarrow \theta \neq 0, \pm\pi, \pm 2\pi, \ldots$

$\Rightarrow \theta \neq n\pi$, where $n \in Z$.

$\therefore$ The domain of the cosecant function $= \mathbb{R} - \{x | x = n\pi,\ n \in Z\}$

As $\csc\theta$ attains all values except those between -1 and 1.

$\therefore$ The range of the cosecant function $= \mathbb{R} - \{x | -1 < x < 1\}$.

The following table summarizes the domains and ranges of the trigonometric functions:

Function	Domain	Range
$y = \sin x$	$-\infty < x < +\infty$	$-1 \le y \le 1$
$y = \cos x$	$-\infty < x < +\infty$	$-1 \le y \le 1$
$y = \tan x$	$-\infty < x < +\infty, x \neq \frac{2n+1}{2}\pi, n \in Z$	$-\infty < y < +\infty$
$y = \cot x$	$-\infty < x < +\infty, x \neq n\pi, n \in Z$	$-\infty < y < +\infty$
$y = sec x$	$-\infty < x < +\infty, x \neq \frac{2n+1}{2}\pi, n \in Z$	$y \ge 1 or\ y \le -1$
$y = \csc x$	$-\infty < x < +\infty, x \neq n\pi, \quad n \in Z$	$y \ge 1 or\ y \le -1$

PERIOD OF TRIGONOMETRIC FUNCTIONS

All the six trigonometric functions repeat their values for each increase or decrease of 2π in θ, i.e., the values of the trigonometric functions for θ and $\theta \pm 2n\pi$, where $\theta \in \mathbb{R}$ and $n \in Z$, are the same. This behavior of the trigonometric functions is called **periodicity.**

The **period** of a trigonometric function is the smallest $+ve$ number which, when added to the original circular measure of the angle, gives the same value of the function.

Let us now discover the periods of the trigonometric functions.

Sine is a periodic function and its period is 2π.

Proof: Suppose p is the period of the sine function such that

$\sin(\theta + p) = \sin\theta$ for all $\theta \in \mathbb{R}$ (1)

Now put $\theta = 0$, we have

$$\sin(0 + p) = \sin 0$$

$$\Rightarrow \sin p = 0$$

$\Rightarrow \quad p = 0, \pm\pi, \pm 2\pi, \pm 3\pi, \ldots$

i) If $p = \pi$, then from (1)

$\sin(\theta + \pi) = \sin\theta$ (not true)

$\therefore \sin(\theta + \pi) = -\sin\theta$

$\therefore \pi$ is not the period of $\sin\theta$.

ii) If $p = 2\pi$, then from (1)

$\sin(\theta + 2\pi) = \sin\theta$, which is true.

As 2π is the smallest $+ve$ real number for which

$\sin(\theta + 2\pi) = \sin\theta$

$\therefore 2\pi$ is the period of $\sin\theta$.

Tangent is a periodic function and its period is π.

Proof: Suppose p is the period of the tangent function such that:

$$\tan(\theta + p) = \tan\theta \qquad \text{for all } \theta \in \mathbb{R} \tag{2}$$

Now put $\theta = 0$, we have

$$\tan(0 + p) = \tan 0$$

$$\Rightarrow \tan p = 0$$

$\therefore\ p = 0, \pi, 2\pi, 3\pi, \ldots$

i) If $p = \pi$, then from (2):

$\tan(\theta + \pi) = \tan\theta$, which is true.

As π is the smallest $+ve$ number for which

$\tan(\theta + \pi) = \tan\theta$

$\therefore\ \pi$ is the period of $\tan\theta$.

By adopting the procedure used in finding the periods of the sine and tangent, we can prove that:

i) 2π is the period of $\cos\theta$ ii) 2π is the period of $\csc\theta$

iii) 2π is the period of $\sec\theta$ iv) π is the peiod of $\cot\theta$.

Example: Find the periods of the following functions:

$\sin 3x$

$\tan 4x$

$\csc \dfrac{x}{4}$

$\cot 8x$

$3\cos\dfrac{x}{5}$

1. $\sec 9x$.

Solution: To find the period we add general periods to all trigonometric functions. For $\sin x, \cos x, \sec x, \csc x$ we will add 2π, and for $\tan x, \cot x$, we will add π.

1. $\sin 3x$

$$= \sin(3x + 2\pi)$$

$$= \sin 3\left(x + \frac{2\pi}{3}\right)$$

So $\frac{2\pi}{3}$ is period of $\sin 3x$.

2. $\tan 4x$

$$= \tan(4x + \pi)$$

$$= \tan 4\left(x + \frac{\pi}{4}\right)$$

So $\frac{\pi}{4}$ is period of $\tan 4x$.

3. $\csc \frac{x}{4}$

$$= \csc\left(\frac{x}{4} + 2\pi\right)$$

$$= \csc\frac{1}{4}\left(x + \frac{2\pi}{\frac{1}{4}}\right)$$

$$= \csc\frac{1}{4}(x + 8\pi)$$

So 8π is period of $\csc\frac{x}{4}$.

4. $\cot 8x$

$= \tan(8x + \pi)$

$= \tan 8\left(x + \frac{\pi}{8}\right)$

So $\frac{\pi}{8}$ is the period of $\cot 8x$.

5. $3\cos\frac{x}{5}$

$= 3\cos\left(\frac{x}{5} + 2\pi\right)$

$= 3\cos\frac{1}{5}(x + 10\pi)$

So 10π is the period of $3\cos\frac{x}{5}$.

6. $\sec 9x$

$= \sec(9x + 2\pi)$

$= \sec 9(x + 18\pi)$

So 18π is the period of $\sec 9x$.

VALUES OF TRIGONOMETRIC FUNCTIONS

We know the values of the trigonometric functions for angles $0°, 30°, 45°$, $60°$, and $90°$. We have also established the following identities:

$\sin(-\theta) = -\sin\theta$ $\cos(-\theta) = \cos\theta$ $\tan(-\theta) = -\tan\theta$

$\sin(\pi - \theta) = \sin\theta$	$\cos(\pi - \theta) = -\cos\theta$	$\tan(\pi - \theta) = -\tan\theta$
$\sin(\pi + \theta) = -\sin\theta$	$\cos(\pi + \theta) = -\cos\theta$	$\tan(\pi + \theta) = \tan\theta$

$\sin(2\pi-\theta)=-\sin\theta$	$cos(2\pi-\theta)=\cos\theta$	$tan(2\pi-\theta)=-\tan\theta$
$\sin(2\pi+\theta)=\sin\theta$	$\cos(2\pi+\theta)=\cos\theta$	$\tan(2\pi+\theta)=\tan\theta$

Using the above identities, we can easily find the values of the trigonometric functions of the angles:

$$-30°,-45°,-60°,-90°, \pm120°, \pm135°, \pm180° \pm210°, \pm225°, \pm240°, \pm270°, \pm300°, \pm315°, \pm330°, \pm360° \cdot$$

GRAPHS OF TRIGONOMETRIC FUNCTIONS

We shall now learn the method of drawing graphs for all six trigonometric functions. These graphs are used very often in calculus and sciences. In order to graph the linear equations of the form:

$$a_1x+b_1y+c_1=0 \quad \text{(i)}$$

$$a_2x+b_2y+c_2=0 \quad \text{(ii)}$$

We remind ourselves of the following:

i) Tables of ordered pairs are constructed from the given equations.

ii) The points corresponding to these ordered pairs are plotted/located.

iii) The points are joined by line segments.

Now, exactly the same procedure is adopted when drawing graphs of the trigonometric functions, with the exception of joining the points with line segments.

For this purpose:

i) A table of ordered pairs (x, y) is constructed, with x as the measure of the angle and y as the value of the trigonometric ratio for the angle of measure x.

ii) The angles are taken along the $\text{x}-$axis.

iv) The values of the trigonometric functions are taken along the $\text{Y}-$axis.

v) These points are joined with the help of smooth curves.

Note: As we shall see, the graphs of trigonometric functions will be smooth curves with no line segments sharp corners or breaks within their domains. This behavior of the curve is called continuity, meaning that the trigonometric functions are continuous, wherever they are defined. Moreover, as the trigonometric functions are periodic, their curves repeat after fixed intervals.

HORIZONTAL AND VERTICAL SHIFTS

The graph of a trigonometric function can be shifted vertically by adding a nonzero constant to the function and horizontally by adding a nonzero constant to the angle of the trigonometric function. If c is a positive number, then adding it to a trigonometric function results in the graph being shifted up c units and subtracting it from a trigonometric function results in the graph being shifted down c units.

For a positive number d, a trigonometric function is shifted left d units when d is added to the angle and shifted right d units when d is subtracted from the angle.

Graph of $y = sinx$ from -2π to 2π

We know that the period of the sine function is 2π, hence, we will first draw the graph for the interval from $0°$ to $360°$ i.e., from 0 to 2π.

To graph the sine function, first recall that $-1 \leq \sin x \leq 1$ for all xò$\mathbb{R}$,

i.e., the range of the sine function is $[-1,1]$, so the graph will be between the horizontal lines $y = 1$ and $y = -1$.

The table of the ordered pairs satisfying $y = \sin x$ is as follows:

x	0	30°	60°	90°	120°	150°	180°	210°	240°	270°	300°	330°	360°
$\sin x$	0	0.5	0.87	1	0.87	0.5	0	-0.5	-0.87	-1	- 0.87	-0.5	0

Now to draw the graph:

- Take a convenient scale $\begin{cases} 1\, side\, of\ small\, square\, on\, the\, x-axis = 10^\circ \\ 1\, side\, of\ small\, square\, on\, the\, y-axis = 1\, unit \end{cases}$
- Draw the coordinate axes.
- Plot the points corresponding to the ordered pairs in the table above i.e., $(0,0),(30^\circ,0.5),(60^\circ,0.87)$ and so on.
- Join the points with the help of a smooth curve, as shown in figure, so we get the graph of $y = \sin x$ from $0\,\text{to}\,360^\circ$, i.e., from $0\,\text{to}\,2\pi$.

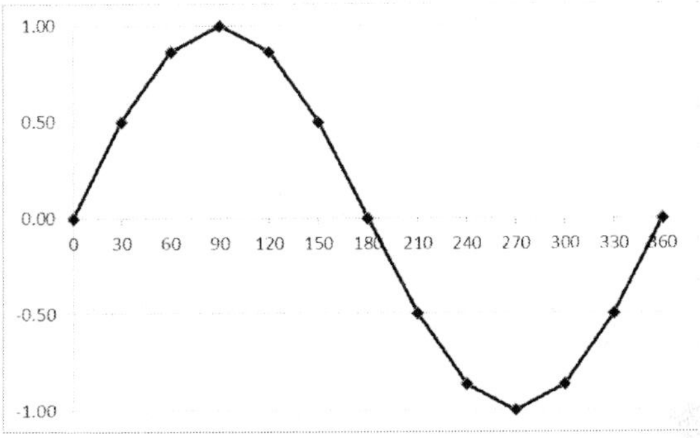

Graph of from.

In a similar way, we can draw the graph for the interval from -360° to 0°. This will complete the graph of $y = \sin x$ from -360° to 360°, i.e., from $-2\pi\ to\ 2\pi$. Which is given as below:

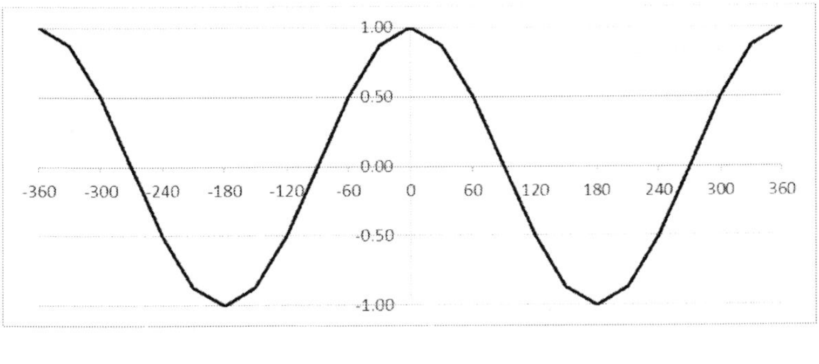

Graph of $y = sinx$from -360° to 360°.

The graph in the interval $[0\ ,2\pi]$ is called a cycle. Since the period of the sine function is 2π, the sine graph can be extended on both sides of the $x-$axis through every interval of 2π.

Graph of $y = \cos x$ from $-2ð\ to 2ð$

We know that the period of the cosine function is 2π, hence, we will first draw the graph for the interval from $0°$ to $360°$, i.e., from $0\ to\ 2\pi$.

To graph the cosine function, first recall that $-1 \le \cos x \le 1$ for all $xò\mathbb{R}$.

i.e., the range of the cosine function is $[-1,1]$, so the graph will be between the horizontal lines $y = 1$ and $y = -1$.

The table of the ordered pairs satisfying $y = \cos x$ is as follows:

x	0	30°	60°	90°	120°	150°	180°	210°	240°	270°	300°	330°	360°
$\cos x$	1	0.87	0.5	0	-0.5	-0.87	-1	-0.87	-0.5	0	0.5	0.87	1

The graph of $y = \cos x$ from $0^0\ to\ 360^0$ is given below:

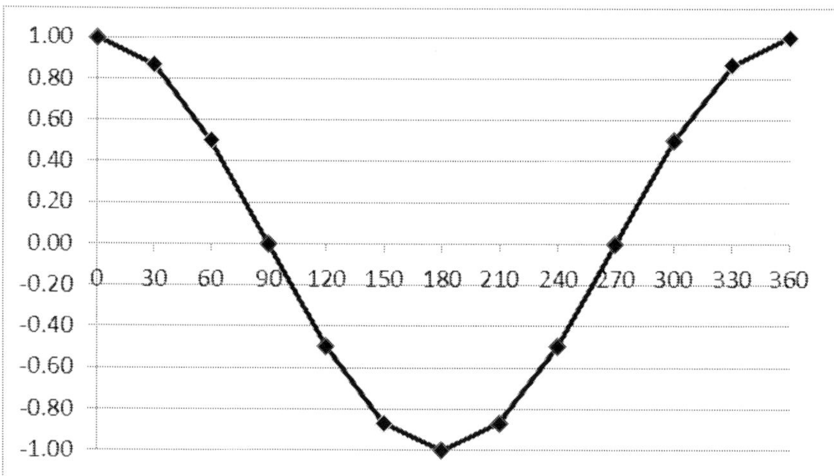

Graph of $y = cosx$ from $0°$ to $360°$.

In a similar way, we can draw the graph for the interval from -360° to 0°. This will complete the graph of $y = \cos x$ from -360° to 360° i.e., from -2π to 2π, which is given as below:

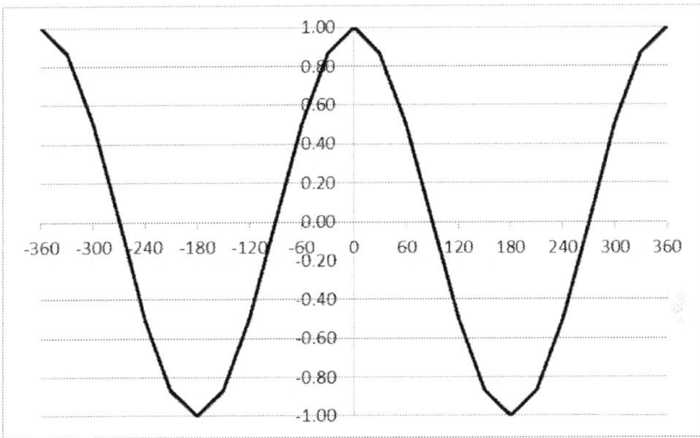

Graph of $y = \cos x$ from -360° to 360°.

The graph in the interval $[0, 2\pi]$ is called a cycle. Since the period of the cosine function is 2π, the cosine graph can be extended on both sides of the x – axis through every interval of 2π.

Graph of $y = \tan x$ from $-\pi$ to π

We know that $\tan(-x) = -\tan x$ and $\tan(\pi - x) = -\tan x$, so the values of $\tan x$ for $x = 0^\circ, 30^\circ, 45^\circ, 60^\circ$ can help us create the table. Also, we know that $\tan x$ is undefined at $x = \pm 90^\circ$.

i) When x approaches $\frac{\pi}{2}$ from the left, i.e., $x \to \frac{\pi}{2} - 0$, $\tan x$ increases indefinitely in the quadrant I.

ii) When x approaches $\frac{\pi}{2}$ from the right, i.e., $x \to \frac{\pi}{2} + 0$, $\tan x$ increases indefinitely in quadrant IV.

iii) When x approaches $-\frac{\pi}{2}$ from the left, i.e., $x \to -\frac{\pi}{2} - 0$, $\tan x$ increases indefinitely in quadrant II.

iv) When x approaches $-\frac{\pi}{2}$ from the right, i.e., $x \to -\frac{\pi}{2}+0,$ $\tan x$ increases indefinitely in quadrant III.

We know that the period of the tangent function is π, so we shall first draw the graph for the interval from $-\pi \; to \; \pi$ i.e., from -180° to 180°.

The table of ordered pairs satisfying $y = tanx$ is given below:

x	-180°	-150°	-120°	-90°	-60°	-30°	0°	30°	60°	90°	120°	150°	180°
$\tan x$	0	0.58	1.73	$+\infty$	-1.73	-0.58	0	0.58	1.73	∞	-1.73	-0.58	0

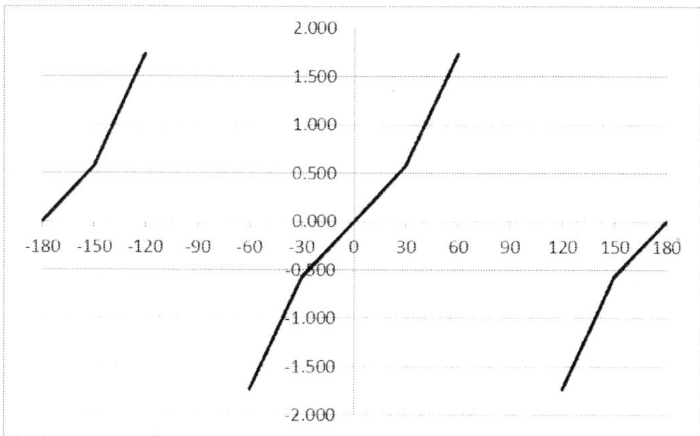

Graph of $y = tan\, x$ from -180° to 180°.

We know that the period of the tangent function is π. The graph is extended on both sides of the x – axis through an interval of π in the same pattern, and so we obtain the graph of $y = \tan x$ from -360° to 360°.

Graph of $y = cot\, x$ from –ð to ð

We know that $\cot(-x) = -cotx$ and $\cot(\pi - x) = -\cot x,$ hence, the values of $\cot x$ for $x = 0^\circ, 30^\circ, 45^\circ, 60^\circ$ can help us in making the table.

Since the period of the cotangent function is also π, its graph is drawn in a similar way to that of the tangent graph using the table given below for the interval -180° to 180°.

x	-180°	-150°	-120°	-90°	-60°	-30°	0°	30°	60°	90°	120°	150°	180°
$\cot x$	$\pm\infty$	01.73	0.58	$+\infty$	-0.58	-1.73	$\pm\infty$	1.73	0.58	$+\infty$	-0.58	-1.73	$\pm\infty$

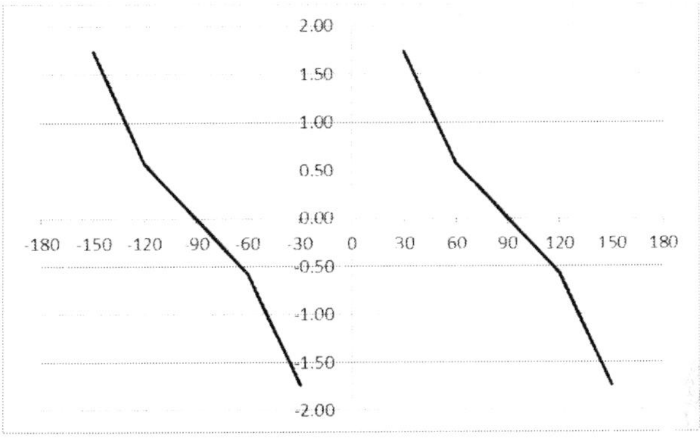

Graph of $y = cotx$from -180° to180°.

Graph of $y = sec\, x$ from $-2\pi$to 2π

We know that $\sec(-x) = \sec x$ and $\sec(\pi - x) = -\sec x$, hence, the values of $\sec x$ for

$0^\circ, 30^\circ, 45^\circ, 60^\circ$ can help us in making the following table of the ordered pairs for the graph of $y = \sec x$ over the interval 0^0 to 360^0.

x	0	30°	60°	90°	120°	150°	180°	210°	240°	270°	300°	330°	360°
$\sec x$	1	1.15	2	∞	-2	-1.15	-1	-1.15	-2	∞	2	1.15	1

Since the period of $\sec x$ is also 2π, we have the following graph of $y = \sec x$ from -360° to360°,. i.e., from -2π to2π.

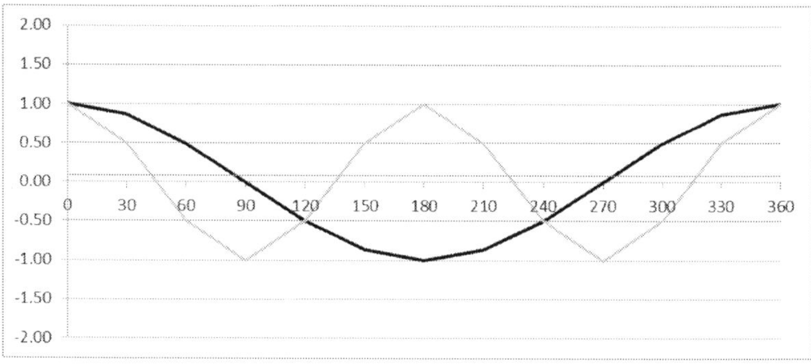

Review of Common Terms and Basic Formulas

- Table of domain and range of trigonometric functions

Function	Domain	Range
$y = \sin x$.	$-\infty < x < +\infty$	$-1 \le y \le 1$
$y = \cos x$.	$-\infty < x < +\infty$.	$-1 \le y \le 1$.
$y = \tan x$	$-\infty < x < +\infty, x \neq \frac{2n+1}{2}\pi, n \in Z$.	$-\infty < y < +\infty$.
	$-\infty < x < +\infty, x \neq n\pi, n \in Z$	$-\infty < y < +\infty$.
		$y \ge 1$ *or* $y \le -1$.
		$y \ge 1$ *or* $y \le -1$.

- All the six trigonometric functions repeat their values for each increase or decrease of 2π. in θ,. i.e., the values of trigonometric functions for θ and $\theta \pm 2n\pi$,. where $\theta \in \mathbb{R}$. and $n \in Z$,. are the same. This behavior of trigonometric function is called **periodicity.**
- The **Period** of a trigonometric function is the smallest $+ve$ number which, when added to the original circular measure of the angle, gives the same value of the fn.

i is the period of $\sin\theta$.

ii) 2π is the period of $\csc\theta$.

i) 2π is the period of $\cos\theta$.

iv) 2π is the peiod of $\sec\theta$.

v) π is the period of $\tan\theta$.

vi) π is the period of $\cot\theta$. .

Steps to draw a graph of trigonometric functions:

i) Table of ordered pairs (x, y) is constructed, where x is the measure of the angle and y is the value of the trigonometric ratio for the angle of measure x.

ii) The angles are tak along the $\text{x}-$axis.

iv) The values of the trigonometric functions are taken along the $\text{y}-$axis.

v) These points are jned with the help of smooth curves.

- The graphs of trigonometric functions will be smooth curves, with no line segments or sharp corners or breaks within their domains. This behavior of the curve is called continuity, meaning that the trigonometric functions are continuous, wherever they are defined. Moreover, as the trigonometric functions are periodic, their curves repeat after fixed intervals. Below are the graphs of all trigonometric functions.

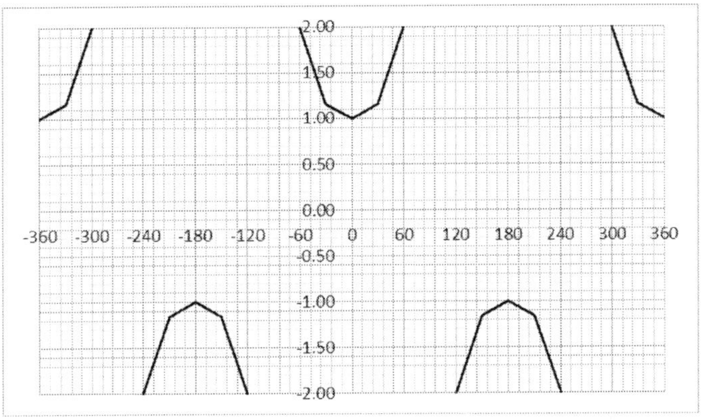

Graph of $y = sec\,x$from -360° to 360°.

MULTIPLE GRAPHS ON THE SAME AXIS

If we need more than one graph on the same axis, we can draw them with the same procedure:

- making tables for both functions, with ordered pairs that satisfy the function, and then plotting these on the same axis. Here are some examples of multiple graphs on the same axis:

Example 1: Draw the graph of the following function for the given intervals:

$y = \sin x$ and $= \sin 2x \ x \epsilon \left[0, 2\pi\right]$.

Solution: To draw the graph of the functions mentioned above, first we will make a table and according to that table we will draw the graphs.

The table of ordered pairs satisfying $y = \sin x$ and $y = \sin 2x$ is given below:

x	0	30°	60°	90°	120°	150°	180°	210°	240°	270°	300°	330°	360°
$y_1 =$	0	0.5	0.87	1	0.87	0.5	0	-0.5	-0.87	-1	-0.87	-0.5	0
$y_1 =$	0	0.87	0.87	0	-0.87	-0.87	0	0.87	0.87	0	-0.87	-0.87	0

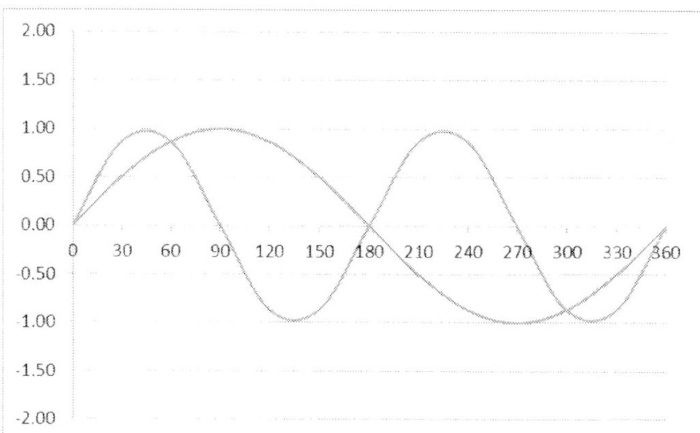

$\sin x$ *Blue line* .

$y_2 = \sin 2x$ *Red line*

Example 2: Draw the graph of the following function for the given intervals.

$y = \cos x$. and $= \cos 2x$. $x \epsilon [0, 2\pi]$.

Solution: To draw the graph of the functions mentioned above, first we will make a table and the according to that table we will draw the graph.

T table of the ordered pairs satisfying $y = \cos x$ and $y = \cos 2x$ is given below:

		30°	60°	90°	120°	150°.	180°.	210°	240°.	270°	300°	330°.	360°.
$y_1 = \cos x$.	1	0.87	0.5	0	-0.5	-0.87	-1	-0.87	-0.5	0	0.5	0.87	1
$y_1 = \cos 2x$.	1	0.5	-0.5	-1	-0.5	0.5	1	0.5	-0.5	-1	-0.5	0.5	1

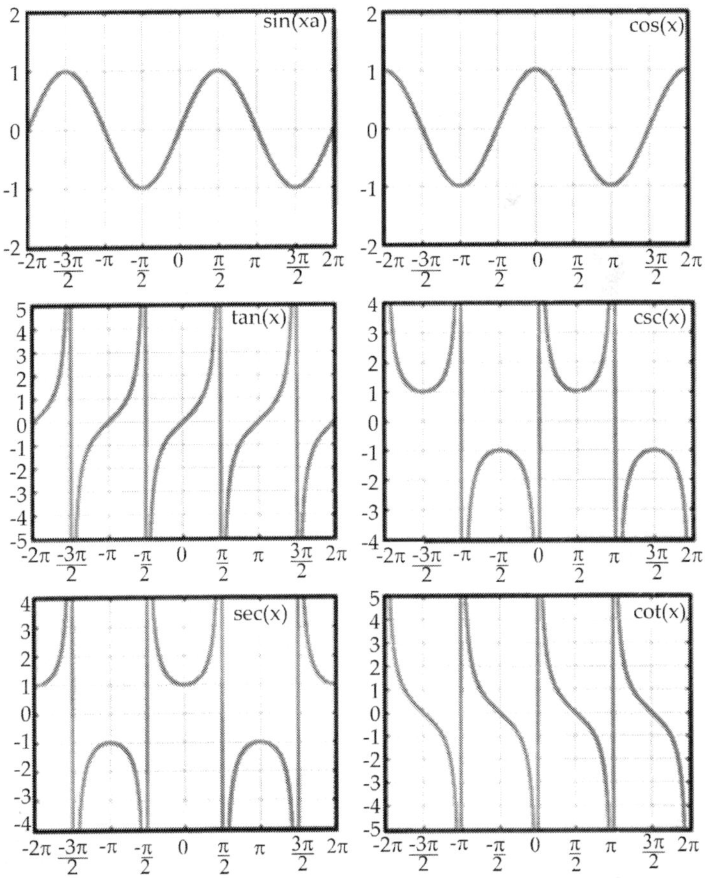

CHAPTER 9

SOLUTIONS OF TRIGONOMETRIC EQUATIONS

INTRODUCTION

Trigonometric equations are, as the name implies, equations that involve trigonometric functions. Similar in many ways to solving polynomial equations or rational equations, only specific values of the variables will be solutions, if there are solutions at all. Often, we will solve a trigonometric equation over a specified interval. However, just as often, we will be asked to find all possible solutions, and as trigonometric functions are periodic, solutions are repeated within each period. In other words, trigonometric equations may have an infinite number of solutions. Additionally, like rational equations, the domain of the function must be considered before we assume that any solution is valid. The period of both the sine function and the cosine function is 2π. In other words, for every 2π, the y-values repeat. If we need to find all possible solutions, then we must add $2\pi k$, where k is an integer, to the initial solution.

Trigonometric equations, i.e., equations involving trigonometric functions of unknown angles, are known as:

(a) Identical equations, or identities, if they are satisfied by all values of the unknown angles for which the functions are defined.

(b) Conditional equations, or equations, if they are satisfied only by particular values of the unknown angles.

First Degree Trigonometric Equations

A trigonometric equation is an equation whose variable is expressed in terms of a trigonometric function value. To solve a quadratic equation, we use the same procedure that we used to solve algebraic equations. For example, the following are examples of trigonometric equations.

$$\sin x = \frac{1}{2}, \sec x = \tan x \text{and} \sin^2 x - \sec x + 1 = \frac{3}{4}.$$

Trigonometric equations have an infinite number of solutions due to the periodicity of the trigonometric functions. For example:

If $\sin\theta = 0$ then $\theta = 0, \pm\pi, \pm 2\pi, \ldots,$

which can be written as $\theta = n\pi, \text{where}\, n \in \mathbb{Z}.$

Now we know that $\sin 30^\circ = \frac{1}{2}$, hence, one value of θ is 30°. We also know that since $\sin\theta$ is positive in the 2nd quadrant, there will be another angle in same quadrant which satisfies the above equation. The relationship between an angle in any quadrant to the acute angle is called the reference angle. The following table shows the relationship of an angle with its reference angle.

Quadrant and Angle	Reference Angle
1st $0 < \theta < 90^\circ$	θ
2nd $90 < \theta < 180^\circ$	$\pi - \theta$
3rd $180 < \theta < 270^\circ$	$-\pi + \theta \, or \, \pi + \theta$
4th $270 < \theta < 360^\circ$	$2\pi - \theta \, or - \theta$

So according to the above example, the second angle will be from the 2nd quadrant, i.e.,

$180^\circ - 30^\circ = 150^\circ$.

Hence, for $sin\theta = \frac{1}{2}$, the solution set is $\{30°, 150°\}$, or in radian form, we can write $\left\{\frac{\pi}{6}, \frac{5\pi}{6}\right\}$.

Procedure to Solve a Trigonometric Equation

1. Solve the equation for the fraction value of the variable.
2. Use a table or calculator to write the exact value of the function.
3. If the angle, as calculated in the 2nd step, is not a quadrantal angle, then find its reference angle.
4. Use the reference angle to find the degree measure of each solution in the interval $0 \le \theta \le 360°$.
5. Add $2n\pi$ to the solution found in step 2 and 4, to write all possible solutions.

Example 1. Solve the Equation: $1 + \cos x = 0$.

Solution: $1 + \cos x = 0$

$\Rightarrow \cos x = -1$.

Since $\cos x$ is –ve, there is only one solution, $x = \pi$ in $[0, 2\pi]$, since 2π is the period of $\cos x$.

$\therefore$ The general value of x is $\pi + 2n\pi$, $n \grave{o} \mathbb{Z}$.

Hence, the solution set $= \{\pi + 2n\pi\}$, $n \grave{o} \mathbb{Z}$.

Example 2. Solve the equation $4\cos^2 x - 3 = 0$

Solution: $4\cos^2 x - 3 = 0$

$\Rightarrow \cos^2 x = \frac{3}{4}$

$\Rightarrow \cos x = \pm\frac{\sqrt{3}}{2}$

i. If $\cos x = \frac{\sqrt{3}}{2}$

$\cos x$ is $+ve$ in the I and IV quadrants with the reference angle $x = \frac{\pi}{6}$

$\therefore x = \frac{\pi}{6}$ and $x = 2\pi - \frac{\pi}{6} = \frac{11\pi}{6}$, where $x \in [0, 2\pi]$,
as 2π is the period of $\cos x$.

$\therefore$ General calues of x are: $\frac{\pi}{6} + 2n\pi$ and $\frac{11\pi}{6} + 2n\pi$, $n \in \mathbb{Z}$.

ii. If $\cos x = -\frac{\sqrt{3}}{2}$:

Since $\cos x$ is $-ve$ in the II and III quadrants with the reference angle

$x = \frac{\pi}{6}$

$\therefore x = \pi - \frac{\pi}{6} = \frac{5\pi}{6}$ and $x = \pi + \frac{\pi}{6} = \frac{7\pi}{6}$ where $x \in [0, 2\pi]$, as 2π is the period of $\cos x$.

$\therefore$ General values of x are: $\frac{5\pi}{6} + 2n\pi$ and $\frac{7\pi}{6} + 2n\pi$, $n \in \mathbb{Z}$,

Hence, we have the solution set

$$= \left\{\frac{\pi}{6} + 2n\pi\right\} \cup \left\{\frac{11\pi}{6} + 2n\pi\right\} \cup \left\{\frac{5\pi}{6} + 2n\pi\right\} \cup \left\{\frac{7\pi}{6} + 2n\pi\right\}, n \in \mathbb{Z}.$$

SOLUTION OF GENERAL TRIGONOMETRIC EQUATIONS

When a trigonometric equation contains more than one trigonometric function, trigonometric identities and algebraic formulae are used to transform such trigonometric equations to an equivalent equation that contains only one trigonometric function.

Example 3. Solve $\sin x + \cos x = 0$.

Solution: $\sin x + \cos x = 0$

$\Rightarrow \dfrac{\sin x}{\cos x} + \dfrac{\cos x}{\cos x} = 0$

$\Rightarrow \tan x + 1 = 0$

$\Rightarrow \tan x = -1$

$\therefore$ $\tan x$ is –ve in the II and IV quadrants with the reference angle $x = \frac{\pi}{4}$

$\therefore \quad x = \pi - \dfrac{\pi}{4} = \dfrac{3\pi}{4}$, where $x \in [0, \pi]$, as π is the period of $\tan x$,

$\therefore$ General value of x is $\frac{3\pi}{4} + n\pi, \qquad n \in \mathbb{Z}$.

$\therefore$ Solution set $= \left\{ \frac{3\pi}{4} + n\pi \right\}, n \in \mathbb{Z}$.

Example 4. Find the solution set of $\sin x \cos x = \frac{\sqrt{3}}{4}$.

Solution: $\sin x \cos x = \frac{\sqrt{3}}{4}$

$\Rightarrow \frac{1}{2}(2 \sin x \cos x) = \frac{\sqrt{3}}{4}$

$\Rightarrow \sin 2x = \frac{\sqrt{3}}{2}$

$\therefore$ $\sin 2x$ is +ve in the I and II quadrants with the reference angle $2x = \dfrac{\pi}{3}$.

$\therefore \quad 2x = \frac{\pi}{3}$ *and* $2x = \pi - \frac{\pi}{3} = \frac{2\pi}{3}$ are two solutions in $[0, 2\pi]$, as 2π is the period of $\sin 2x$.

$\therefore$ General values of $2x$ are $\frac{\pi}{3}+2n\pi$ and $\frac{2\pi}{3}+2n\pi,\quad n\in\mathbb{Z}$

$\Rightarrow$ General values of $2x$ are $\frac{\pi}{3}+2n\pi$ and $\frac{2\pi}{3}+2n\pi\quad n\in\mathbb{Z}$

Hence, the solution set $=\left\{\frac{\pi}{6}+n\pi\right\}\cup\left\{\frac{\pi}{3}+n\pi\right\}$.

Example 5: Solve the equation $\sin 2x=\cos x$**.**

Solution: $\sin 2x=\cos x$

$\Rightarrow 2\sin x\cos x=\cos x$

$\Rightarrow 2\sin x\cos x-\cos x=0$

$\cos x(2\sin x-1)=0$

$\therefore \cos x=0 \quad or \quad 2\sin x-1=0$.

i. if $\cos x=0$

$\Rightarrow x=\frac{\pi}{2}$ and $x=\frac{3\pi}{2}$, where $x\in[2,2\pi]$, as 2π is the period of $\cos x$.

$\therefore$ General values of x are $\frac{\pi}{2}+2n\pi$ and $\frac{3\pi}{2}+2n\pi,\ n\in\mathbb{Z}$.

ii. If $2\sin x-1=0$

$\Rightarrow \sin x=\frac{1}{2}$.

Since $\sin x$ is +ve in the I and II quadrants with the reference angle $x=\frac{\pi}{6}$,

$\therefore x=\frac{\pi}{6}$ and $x=\pi-\frac{\pi}{6}=\frac{5\pi}{6}$,

where $x\in[0,2\pi]$, as 2π is the period of $\sin x$.

$\therefore$ General values of x are $\frac{\pi}{6}+2n\pi$ and $\frac{5\pi}{6}+2n\pi,\ n\in\mathbb{Z}\cdot$

Hence, the solution set $=\left\{\frac{\pi}{2}+2n\pi\right\}\cup\left\{\frac{3\pi}{2}+2n\pi\right\}\cup\left\{\frac{\pi}{6}+2n\pi\right\}\cup\left\{\frac{5\pi}{6}+2n\pi\right\},\ n\in\mathbb{Z}$

.

Example 6: Solve the equation $\sin^2 x+\cos x=1$.

Solution: $\sin^2 x+\cos x=1$

$\Rightarrow 1-\cos^2 x+\cos x-1=0$

$\Rightarrow \cos x-\cos^2 x=0$

$\cos x\,(1-\cos x)=0$

$\Rightarrow \cos x=0$ or $1-\cos x=0$.

Now, if $\cos x=0$

$\Rightarrow x=\frac{\pi}{2}$ and $x=\frac{3\pi}{2}$ where $x\in[0,2\pi]$, as 2π is the period of $\cos x$.

$\therefore$ General values of $x=\frac{\pi}{2}+2n\pi$ and $\frac{3\pi}{2}+2n\pi,\qquad n\in\mathbb{Z}.$

If $\cos x=1$,

$\Rightarrow x=0$ and $x=2\pi$, where $x\in[0,2\pi]$, as 2π is the period of $\cos x$.

$\therefore$ General values of x are $0+2n\pi\cup 2\pi+2n\pi, n\in\mathbb{Z}$.

$\therefore$ Solution Set $=\left\{\frac{\pi}{2}+2n\pi\right\}\cup\left\{\frac{3\pi}{2}+2n\pi\right\}\cup\{2n\pi\}\cup\{2\pi+2n\pi\}, n\in\mathbb{Z}\cdot$

$\therefore$ Since $\{2(n+1)\pi\}\subset\{2n\pi\}, n\in\mathbb{Z}$.

Hence, the solution set $=\left\{\frac{\pi}{2}+2n\pi\right\}\cup\left\{\frac{3\pi}{2}+2n\pi\right\}\cup\{2n\pi\}, n\in\mathbb{Z}.$

***Extraneous Roots*:** Sometimes it is necessary to square both sides of a trigonometric equation. In such a case, extraneous roots can occur, which are to be discarded. Hence, each value of x must be checked by substituting it into the given equation.

For example, $x = 2$ is an equation having a root of 2. On squaring, we get $x^2 = 4$, which gives two roots 2 and –2. However, the root –2 does not satisfy the equation $x = 2$. Therefore, –2 is an extraneous root.

Example 7: $\tan^2\theta - \sec\theta - 1 = 0$.

Solution: $\tan^2\theta - \sec\theta - 1 = 0$

$\Rightarrow (\sec^2\theta - 1) - \sec\theta - 1 = 0$

$\Rightarrow \sec^2\theta - \sec\theta - 2 = 0$

$\Rightarrow \sec^2\theta - 2\sec\theta + \sec\theta - 2 = 0$

$\Rightarrow \sec\theta(\sec\theta - 2) + 1(\sec\theta - 2) = 0$

$\Rightarrow (\sec\theta - 2)(\sec\theta + 1) = 0$

$\Rightarrow (\sec\theta + 1) = 0$ or $(\sec\theta - 2) = 0$

$\Rightarrow \sec\theta = -1$ or $\sec\theta = 2$.

i. Consider $\sec\theta = -1$

$\Rightarrow \cos\theta = -1$.

As $\cos\theta$ is –ve in the 2nd and 3rd quadrants,

$\Rightarrow$ refrence angle $= \cos^{-1}(1) = 0$.

Since it is a quadrantal angle, we get the same answer for the 2nd and 3rd quadrants.

For the 2nd quadrant: $\pi - \theta = \pi - 0 = \pi$.

For the 3rd quadrant: $\pi + \theta = \pi$.

ii. Now consider $\sec\theta = 2$

$\Rightarrow \cos\theta = \frac{1}{2},$

as $\cos\theta$ is +ve in the 1st and 4th quadrant. Hence, reference angle:

$\cos^{-1}\left(\frac{1}{2}\right) = \frac{\pi}{6}.$

For 1st quadrant: $\frac{\pi}{6}$.

For 4th quadrant: $2\pi - \theta = 2\pi - \frac{\pi}{6} = \frac{11\pi}{6}$, as the period of $\cos\theta$ is 2π. Hence, we will add $2n\pi$ to all possible solutions.

Solution Set = $\{\pi + 2n\pi\} \cup \left\{\frac{\pi}{6} + 2n\pi\right\} \cup \left\{\frac{11\pi}{6} + 2n\pi\right\}$.

Example 8. $3\tan^2 \boldsymbol{x} + 2\sqrt{3}\tan \boldsymbol{x} + 1 = 0$

Solution: $3\tan^2 x + 2\sqrt{3}\tan x + 1 = 0$

$\Rightarrow \left(\sqrt{3}\tan x\right)^2 + 2\left(\sqrt{3}\tan x\right)(1) + (1)^2 = 0$

$\Rightarrow \left(\sqrt{3}\tan x + 1\right)^2 = 0$

$\Rightarrow \left(\sqrt{3}\tan x + 1\right) = 0$

$\Rightarrow \sqrt{3}\tan x = -1$

$\Rightarrow \tan x = -\frac{1}{\sqrt{3}},$

as $\tan x$ is –ve in the 2nd and 4th quadrants, hence:

$\Rightarrow$ refrence angle $= \tan^{-1}\left(\frac{1}{\sqrt{3}}\right) = \frac{\pi}{6}$.

For 2nd quadrant: $\pi - \frac{\pi}{6} = \frac{5\pi}{6}$.

For 4th quadrant: $2\pi - \theta = 2\pi - \frac{\pi}{6} = \frac{11\pi}{6}$,

since the period of $\tan x$ is π. For a solution set, we add $n\pi$ to all possible solutions.

Hence, the solution set = $\left\{\frac{\pi}{6} + n\pi\right\} \cup \left\{\frac{11\pi}{6} + n\pi\right\}$.

Example 9. $\sin \boldsymbol{x} + \sin 3\boldsymbol{x} + \sin 5\boldsymbol{x} = 0$

Solution: $(\sin 5x + \sin x) + \sin 3x = 0$.

Apply the formula of the sum to product conversion:

$\Rightarrow 2\sin(\frac{5x+x}{2}).\cos(\frac{5x-x}{2}) + \sin 3x = 0$

$\Rightarrow 2\sin 3x \cos 2x + \sin 3x = 0$

$\Rightarrow \sin 3x(2\cos 2x + 1) = 0$

$\Rightarrow \sin 3x = 0$ or $2\cos 2x + 1 = 0$.

i. If $\sin 3x = 0$

$\Rightarrow 3x = \sin^{-1}(0)$

$\Rightarrow 3x = 0, \pi$

$\Rightarrow x = 0 + \frac{2n\pi}{3}, \frac{\pi}{3} + \frac{2n\pi}{3}$

$\Rightarrow x = \frac{2n\pi}{3}, (2n+1)\frac{\pi}{3}$.

ii. If $2\cos 2x + 1 = 0$

$$\Rightarrow \cos 2x = -\frac{1}{2}$$

$$\Rightarrow 2x = \cos^{-1}\left(\frac{1}{2}\right)$$

$$\Rightarrow 2x = \frac{2\pi}{3}, \frac{4\pi}{3}$$

$$\Rightarrow x = \frac{\pi}{3}, \frac{2\pi}{3}$$

$$\Rightarrow x = \frac{\pi}{3} + 2n\pi, \frac{2\pi}{3} + 2n\pi \cdot$$

Solution Set $= \left\{\frac{2n\pi}{3}\right\} \cup \left\{\frac{\pi}{3} + \frac{2n\pi}{3}\right\} \cup \left\{\frac{\pi}{3} + n\pi\right\} \cup \left\{\frac{2\pi}{3} + n\pi\right\}.$

Example 10. $3\cos^2 \boldsymbol{x} - 2\sqrt{3}\sin \boldsymbol{x}\cos \boldsymbol{x} - 3\sin^2 \boldsymbol{x} = 0$

Dividing throughout by $\cos^2 x$:

$$3\frac{\cos^2 x}{\cos^2 x} - \frac{2\sqrt{3}(\sin x \cos x)}{\cos^2 x} - 3\frac{\sin^2 x}{\cos^2 x} = 0$$

$$\Rightarrow 3 - 2\sqrt{3}\tan x - 3\tan^2 x = 0$$

$$\Rightarrow -3\tan^2 x - 2\sqrt{3}\tan x + 3 = 0$$

$$\Rightarrow 3\tan^2 x + 2\sqrt{3}\tan x - 3 = 0.$$

Since it is a quadratic equation, we can apply the quadratic formula.

$$\tan x = \frac{-2\sqrt{3} \pm \sqrt{\left(2\sqrt{3}\right)^2 - 4(3)(-3)}}{2(3)}$$

$$\tan x = \frac{\left(-2\sqrt{3} \pm \sqrt{48}\right)}{6}$$

$$\tan x = \frac{\left(-2\sqrt{3} \pm 4\sqrt{3}\right)}{6}$$

$\tan x = \frac{-2\sqrt{3}+4\sqrt{3}}{6}$ or $\tan x = \frac{\left(-2\sqrt{3}-4\sqrt{3}\right)}{6}$

$\tan x = \frac{1}{\sqrt{3}}$ or $\tan x = -\sqrt{3}$

i. When $\tan x = \frac{1}{\sqrt{3}}$.

Since $\tan x$ is $+ve$ is in the 1st and 3rd quadrant:

Reference angle: $\tan^{-1}\left(\frac{1}{\sqrt{3}}\right) = \frac{\pi}{6}$.

In the 1st quadrant: $\theta = \frac{\pi}{6}$.

In the 3rd quadrant: $\pi + \theta = \pi + \frac{\pi}{6} = \frac{7\pi}{6}$.

For the general solution, we will add $n\pi$ to possible solutions, as the period of $\tan x$ is π.

Hence, the general solution is $\left\{\frac{\pi}{6} + n\pi\right\} \cup \left\{\frac{7\pi}{6} + n\pi\right\}$.

ii. When $\tan x = -\sqrt{3}$.

Since $\tan x$ is $-ve$ is in the 2nd and 4th quadrants.

Reference angle: $\tan^{-1}\left(\sqrt{3}\right) = \frac{\pi}{3}$.

In the 2nd quadrant: $\pi - \theta = \pi - \frac{\pi}{3} = \frac{2\pi}{3}$.

In the 4th quadrant: $2\pi - \theta = 2\pi - \frac{\pi}{3} = \frac{5\pi}{3}$.

For a general solution, we will add $n\pi$ to the possible solution, as the period of $\tan x$ is π.

Hence, the general solution is $\left\{\frac{2\pi}{3}+n\pi\right\}\cup\left\{\frac{5\pi}{3}+n\pi\right\}$.

EXTRANEOUS ROOTS PROBLEMS

Example 1. $\sqrt{3}\tan \boldsymbol{x}-\sec \boldsymbol{x}-1=0$

Solution: $\sqrt{3}\tan x-\sec x-1=0$ (1)

$\sqrt{3}\frac{\sin x}{\cos x}-\frac{1}{\cos x}-1=0$

$\sqrt{3}\sin x-1=\cos x$.

On squaring both sides,

$\left(\sqrt{3}\sin x-1\right)^2=(\cos x)^2$

$\Rightarrow 3\sin^2 x+1-2\sqrt{3}\sin x=\cos^2 x$

$\Rightarrow 3\sin^2 x+1-2\sqrt{3}\sin x=1-\sin^2 x$

$\Rightarrow 3\sin^2 x+1-2\sqrt{3}\sin x-1+\sin^2 x=0$

$\Rightarrow 4\sin^2 x-2\sqrt{3}\sin x=0$

$\Rightarrow 2\sin x\left(2\sin x-\sqrt{3}\right)=0$

$\Rightarrow 2\sin x=0$ or $2\sin x-\sqrt{3}=0$

$\Rightarrow \sin x=0$ or $\sin x=\frac{\sqrt{3}}{2}$

$\Rightarrow x=\sin^{-1}(0)$ or $x=\sin^{-1}(\frac{\sqrt{3}}{2})$

$\Rightarrow x = 0, \pi$ or $x = \frac{\pi}{3}, \frac{2\pi}{3}$.

Now to check extraneous root, put all possible values $x = 0,\ \pi,\ \frac{\pi}{3}, \frac{2\pi}{3}$ into (1).

When $x = 0$

$$\text{LH S} = \sqrt{3}\tan(0) - \sec(0) - 1 = -2 \neq 0 = \text{RHS}.$$

This implies that $x = 0$ is an extraneous root of the given equation.

Now let $x = \pi$:

$$\text{LHS} = \sqrt{3}\tan(\pi) - \sec(\pi) - 1 = 0 = \text{RHS}.$$

This implies that $x = \pi$ is a root of the equation.

Now put $x = \frac{\pi}{3}$.

$$\text{LHS} = \sqrt{3}\tan\left(\frac{\pi}{3}\right) - \sec\left(\frac{\pi}{3}\right) - 1 = 0 = \text{RHS}.$$

This implies that $x = \frac{\pi}{3}$ is a root of a given equation.

Put $x = \frac{2\pi}{3}$.

$$\text{LH S} = \sqrt{3}\tan\left(\frac{2\pi}{3}\right) - \sec\left(\frac{2\pi}{3}\right) - 1 = -2 \neq \text{RHS}$$

This implies that $x = \frac{2\pi}{3}$ is an extreanous root.

Since we obtained all results from sine function, and the period of the sine function is 2π, hence, the general solution $=\{\pi+2n\pi\}\cup\left\{\frac{\pi}{3}+2n\pi\right\}$.

Example 2. Solve the equation: $\csc x=\sqrt{3}+\cot x$.

Solution: $\csc x=\sqrt{3}+\cot x$.

$$\Rightarrow \frac{1}{\sin x}=\sqrt{3}+\frac{\cos x}{\sin x}$$

$$\Rightarrow 1=\sqrt{3}\sin x+\cos x$$

$$\Rightarrow 1-\cos x=\sqrt{3}\sin x$$

$$\Rightarrow (1-\cos x)^2=\left(\sqrt{3}\sin x\right)^2$$

$$\Rightarrow 1-2\cos x+\cos^2 x=3\sin^2 x$$

$$\Rightarrow 1-2\cos x+\cos^2 x=3\left(1-\cos^2 x\right)$$

$$\Rightarrow 4\cos^2 x-2\cos x-2=0$$

$$\Rightarrow 2\cos^2 x-\cos x-1=0$$

$$\Rightarrow (2\cos x+1)(\cos x-1)=0$$

$$\Rightarrow \cos x=-\frac{1}{2} \quad \text{or } \cos x=1.$$

i. If $\cos x=-\frac{1}{2}$

Since $\cos x$ is –ve in the II and III quadrants, with the reference angle

$$x=\frac{\pi}{3}$$

$\Rightarrow\ x=\pi-\frac{\pi}{3}=\frac{2\pi}{3}$ and $x=\pi+\frac{\pi}{3}=\frac{4\pi}{3}$ where $x\in[0\ ,2\pi]$.

However, $x=\frac{4\pi}{3}$ does not satisfy the given equation, hence, only $x=\frac{2\pi}{3}$ is the solution.

ii. If $\cos x=1$

$\Rightarrow\ x=0$ and $x=2\pi$ where $x\in[0\ ,2\pi]$.

Now both $\csc x$ and $\cot x$ are not defined for $x=0$ and $x=2\pi$

$\therefore\ x=0$ and $x=2\pi$ are not admissible.

Hence the solution set $=\left\{\frac{2\pi}{3}+2n\pi\right\}$.

Some Trigonometric Equation using the Formulas of Sums and Differences and identities

Example 1. $\cos 2x=\sin 3x$

$\Rightarrow\ \cos^2 x-\sin^2 x=3\sin x-4\sin^3 x$

$\Rightarrow\ \left(1-\sin^2 x\right)-\sin^2 x-3\sin x+4\sin^3 x=0$

$\Rightarrow\ 1-2\sin^2 x-3\sin x+4\sin^3 x=0$

$\Rightarrow\ 4\sin^3 x-2\sin^2 x-3\sin x+1=0$.

$\Rightarrow$ We will solve this cubic equation by synthetic division.

Take $\sin x=1$ as a root, then by synthetic division

1	4	-2	-3	1
↓	4	2	-1	
4	2	-1	0	

$\Rightarrow\ (\sin x-1)(4\sin^2 x+2\sin x-1)=0$

$\Rightarrow\ \sin x-1=0 \quad$ or $4\sin^2 x+2\sin x-1=0$.

i. If $\sin x=1$

$x=\sin^{-1}(1)$

Since it is a quadrantal angle, $x=\frac{\pi}{2}$, and the general solution will be $\left\{\frac{\pi}{2}+2n\pi\right\}$.

ii. If $4\sin^2 x+2\sin x-1=0$, apply the quadratic formula,

$$\sin x=\frac{-2\pm\sqrt{(2)^2-4(4)(-1)}}{2(4)}$$

$$\sin x=\frac{-2\pm\sqrt{4+16}}{2(4)}$$

$$\sin x=\frac{-2\pm\sqrt{20}}{2(4)}$$

$$\sin x=\frac{-2\pm2\sqrt{5}}{8}$$

$$\sin x=\frac{2(-1\pm\sqrt{5})}{8}.$$

Here, we have two cases:

$\sin x = \frac{-1+\sqrt{5}}{4}$ or $\sin x = \frac{-1-\sqrt{5}}{4}$.

a). If $\sin x = \frac{-1+\sqrt{5}}{4}$

$\Rightarrow x = \sin^{-1}(0.309)$

$\Rightarrow x \approx 18^\circ, 162^\circ$.

Converting degrees to radians, we obtain:

$\Rightarrow x = \frac{\pi}{10}, \frac{9\pi}{10}$.

The general solution will be $\left\{\frac{\pi}{10} + 2n\pi\right\} \cup \left\{\frac{9\pi}{10} + 2n\pi\right\}$.

b) If $\sin x = \frac{-1-\sqrt{5}}{4}$

$\Rightarrow x = \sin^{-1}(-0.809)$

$\Rightarrow x \approx 234^\circ, 306^\circ$.

Converting degrees to radians, we obtain:

$\Rightarrow x = \frac{13\pi}{10}, \frac{17\pi}{10}$.

The general solution will be $\left\{\frac{13\pi}{10} + 2n\pi\right\} \cup \left\{\frac{17\pi}{10} + 2n\pi\right\}$, hence the complete solution is $= \left\{\frac{\pi}{10} + 2n\pi\right\} \cup \left\{\frac{9\pi}{10} + 2n\pi\right\} \cup \left\{\frac{13\pi}{10} + 2n\pi\right\} \cup \left\{\frac{17\pi}{10} + 2n\pi\right\}$.

Example 2. $\sec 3\theta = \sec\theta$

$\Rightarrow \frac{1}{\cos 3\theta} = \frac{1}{\cos\theta}$

$\Rightarrow \cos 3\theta = \cos\theta$

$\Rightarrow\ 4\cos^3\theta - 3\cos\theta = \cos\theta$

$\Rightarrow\ 4\cos^3\theta - 4\cos\theta = 0$

$\Rightarrow\ 4\cos\theta\left(\cos^2\theta - 1\right) = 0$

$\Rightarrow\ 4\cos\theta = 0$ Or $\cos^2\theta - 1 = 0$.

i. If $\cos\theta = 0$

$\theta = \cos^{-1}(0)$

$\theta = \frac{\pi}{2}, \frac{3\pi}{2}.$

The general solution is $\left\{\frac{\pi}{2} + 2n\pi\right\} \cup \left\{\frac{3\pi}{2} + 2n\pi\right\}$.

ii. If $\cos^2\theta - 1 = 0$

$\Rightarrow \cos^2\theta = 1$

$\Rightarrow\ cos\theta = \pm 1$

$\Rightarrow\ \cos\theta = 1\ \ or \cos\theta = -1$

$\theta = 0, \pi$ or we can write it as $n\pi$.

Hence, $\theta = n\pi$.

Solution Set $= \left\{\frac{\pi}{2} + 2n\pi\right\} \cup \left\{\frac{3\pi}{2} + 2n\pi\right\} \cup \{n\pi\}$.

Example 3. $\tan 2\theta + \cot\theta = 0$

$\Rightarrow\ \tan 2\theta = -\cot\theta$

$\Rightarrow\ \frac{\sin 2\theta}{\cos 2\theta} = -\frac{\cos\theta}{\sin\theta}$

$$\Rightarrow (\sin 2\theta)(\sin\theta) = -(\cos 2\theta)(\cos\theta)$$

$$\Rightarrow (2\sin\theta\cos\theta)(\sin\theta) = -\cos\theta\left(\cos^2\theta - \sin^2\theta\right)$$

$$\Rightarrow 2\sin^2\theta\cos\theta = -\cos^3\theta + \cos\theta\sin^2\theta$$

$$\Rightarrow \cos^3\theta + 2\sin^2\theta\cos\theta - \sin^2\theta\cos\theta = 0$$

$$\Rightarrow \cos^3\theta + \sin^2\theta\cos\theta = 0$$

$$\Rightarrow \cos\theta\left(\cos^2\theta + \sin^2\theta\right) = 0.$$

As $\cos^2\theta + \sin^2\theta = 1 \neq 0 \quad \Rightarrow \cos\theta = 0$

$$\Rightarrow \theta = \cos^{-1}(0)$$

$$\Rightarrow \theta = \frac{\pi}{2}, \frac{3\pi}{2}.$$

Since the period of $\cos\theta$ is 2π, we will add $2n\pi$ as a general solution.

The general values are $\frac{\pi}{2} + 2n\pi, \frac{3\pi}{2} + 2n\pi$.

Hence, the solution set is $= \left\{\frac{\pi}{2} + 2n\pi\right\} \cup \left\{\frac{3\pi}{2} + 2n\pi\right\}$.

Example 4. $\sin 2x + \sin x = 0$

$$\Rightarrow 2\sin x\cos x + \sin x = 0$$

$$\Rightarrow \sin x(2\cos x + 1) = 0$$

$$\Rightarrow \sin x = 0 \quad \text{or} \quad 2\cos x = -1$$

$$\Rightarrow \sin x = 0 \quad \text{or} \quad \cos x = -\frac{1}{2}.$$

Now if $\sin x = 0$,

$x = n\pi$.

If $\cos x = -\frac{1}{2}$

$x = \frac{2\pi}{3}, \frac{4\pi}{3}$

and the solution set $= \{n\pi\} \cup \left\{\frac{2\pi}{3} + 2n\pi\right\} \cup \left\{\frac{4\pi}{3} + 2n\pi\right\}$.

Example 5. $\sin 4x - \sin 2x = \cos 3x$

$\Rightarrow 2\cos(\frac{4x+2x}{2})\sin\left(\frac{4x-2x}{2}\right) = \cos 3x$

$\Rightarrow \ 2\cos 3x \sin x - \cos 3x = 0$

$\Rightarrow \ \cos 3x(2\sin x - 1) = 0$

$\Rightarrow \ \cos 3x = 0 \qquad or \quad 2\sin x - 1 = 0$.

i. Now if $\cos 3x = 0$

$\Rightarrow \ 3x = \cos^{-1}(0)$

$3x = \frac{\pi}{2}, \frac{3\pi}{2}$.

The general values are:

$\Rightarrow \ 3x = \frac{\pi}{2} + 2n\pi, \qquad 3x = \frac{3\pi}{2} + 2n\pi$

$\Rightarrow \ x = \frac{\pi}{6} + \frac{2n\pi}{3}, \qquad x = \frac{3\pi}{6} + \frac{2n\pi}{3}$

$\Rightarrow \ x = \frac{\pi}{6} + \frac{2n\pi}{3}, \qquad x = \frac{\pi}{2} + \frac{2n\pi}{3}$.

ii. $2\sin x - 1 = 0$

$\Rightarrow 2\sin x = 1$

$\Rightarrow \sin x = \dfrac{1}{2}$

$\Rightarrow x = \dfrac{\pi}{6}, \dfrac{5\pi}{6}.$

The general values are $\left\{\dfrac{\pi}{6} + 2n\pi\right\}, \left\{\dfrac{5\pi}{6} + 2n\pi\right\}.$

Solution set: $\left\{\dfrac{\pi}{6} + \dfrac{2n\pi}{3}\right\} \cup \left\{\dfrac{\pi}{2} + \dfrac{2n\pi}{3}\right\} \cup \left\{\dfrac{\pi}{6} + 2n\pi\right\} \cup \left\{\dfrac{5\pi}{6} + 2n\pi\right\}.$

Example 6. $\sin 7x - \sin x = \sin 3x$

$\Rightarrow 2\cos\left(\dfrac{7x + x}{2}\right)\sin\left(\dfrac{7x - x}{2}\right) = \sin 3x$

$\Rightarrow 2\cos(4x)\sin(3x) - \sin 3x = 0$

$\Rightarrow \sin 3x(2\cos 4x - 1) = 0$

$\Rightarrow \sin 3x = 0$ Or $2\cos 4x - 1 = 0$.

i. If $\sin 3x = 0$

$\Rightarrow 3x = \sin^{-1}(0)$

$\Rightarrow 3x = 0, \pi$

$\Rightarrow 3x = 2n\pi, \pi + 2n\pi$.

The general values are

$\Rightarrow x = \dfrac{2n\pi}{3}, \dfrac{\pi + 2n\pi}{3}$

$$\Rightarrow x = \frac{2n\pi}{3}, (2n+1)\frac{\pi}{3}.$$

ii. If $2\cos 4x - 1 = 0$

$$\Rightarrow \cos 4x = \frac{1}{2}$$

$$\Rightarrow 4x = \cos^{-1}\left(\frac{1}{2}\right)$$

$$\Rightarrow 4x = \frac{\pi}{3}, \frac{5\pi}{3}$$

$$\Rightarrow 4x = \frac{\pi}{3} + 2n\pi, \frac{5\pi}{3} + 2n\pi$$

$$\Rightarrow x = \frac{\pi}{12} + \frac{2n\pi}{4}, \frac{5\pi}{12} + \frac{2n\pi}{4}.$$

The general values are

$$\Rightarrow x = \frac{\pi}{12} + \frac{n\pi}{2}, \frac{5\pi}{12} + \frac{n\pi}{2}.$$

Solution Set $= \left\{\frac{2n\pi}{3}\right\} \cup \left\{(2n+1)\frac{\pi}{3}\right\} \cup \left\{\frac{\pi}{12} + \frac{n\pi}{2}\right\} \cup \left\{\frac{5\pi}{12} + \frac{2n\pi}{4}\right\}.$

Example 7. $\sin\theta + \sin 3\theta + \sin 5\theta + \sin 7\theta = 0$.

$$\Rightarrow (\sin 7\theta + \sin\theta) + (\sin 5\theta + \sin 3\theta) = 0$$

$$\Rightarrow 2\sin\frac{7\theta + \theta}{2}\cos\frac{7\theta - \theta}{2} + 2\sin\frac{5\theta + 3\theta}{2}\cos\frac{5\theta - 3\theta}{2} = 0$$

$$\Rightarrow 2\sin 4\theta\cos 3\theta + 2\sin 4\theta\cos\theta =$$

$$\Rightarrow 2\sin 4\theta(\cos 3\theta + \cos\theta) = 0$$

$$\Rightarrow 2\sin 4\theta\left(2\cos\frac{3\theta+\theta}{2}\cos\frac{3\theta-\theta}{2}\right) = 0$$

$\Rightarrow 4\sin 4\theta \cos 2\theta \cos\theta = 0$

$\sin 4\theta = 0$ or $\cos 2\theta = 0$ or $\cos\theta = 0$.

Now here we have three cases:

i. If $\sin 4\theta = 0$

$\Rightarrow 4\theta = \sin^{-1}(0)$

$\Rightarrow 4\theta = 0, \pi$.

Adding the period of sine, we get:

$\Rightarrow 4\theta = 2n\pi, \pi + 2n\pi$

$\Rightarrow \theta = \frac{2n\pi}{4}, \frac{\pi}{4} + \frac{2n\pi}{4}$.

The general values are:

$\Rightarrow \theta = \frac{n\pi}{2}, (2n+1)\frac{\pi}{4}$.

ii. If $\cos 2\theta = 0$

$\Rightarrow 2\theta = \cos^{-1}(0)$

$\Rightarrow 2\theta = \frac{\pi}{2}, \frac{3\pi}{2}$

$\Rightarrow 2\theta = \frac{\pi}{2} + 2n\pi, \frac{3\pi}{2} + 2n\pi$.

The general Values are

$\Rightarrow \theta = \frac{\pi}{4} + n\pi, \frac{3\pi}{4} + n\pi$.

iii. $\cos\theta = 0$

$\Rightarrow \theta = \cos^{-1}(0)$

$\Rightarrow \theta = \frac{\pi}{2}, \frac{3\pi}{2}.$

The general values are $\theta = \frac{\pi}{2} + 2n\pi, \frac{3\pi}{2} + 2n\pi \cdot$

Hence, the solution set

$$= \left\{\frac{n\pi}{2}\right\} \cup \left\{(2n+1)\frac{\pi}{4}\right\} \cup \left\{\frac{\pi}{4} + n\pi\right\} \cup \left\{\frac{3\pi}{4} + n\pi\right\} \cup \left\{\frac{\pi}{2} + 2n\pi\right\} \cup \left\{\frac{3\pi}{2} + 2n\pi\right\}$$

SUMMARY

After completing this chapter, students will be able to:

- Identify trigonometric equations.
- Solve trigonometric equations.
- Know the general solution of trigonometric equations.
- Understand extraneous roots of trigonometric equations.
- Solve difficult trigonometric equations using different identities.

REVIEW OF COMMON TERMS AND BASIC FORMULAS

Trigonometric equations are, as the name implies, equations that involve trigonometric functions. Trigonometric equations have an infinite number of solutions due to the periodicity of trigonometric functions. For example:

If $\sin\theta = 0$ then $\theta = 0, \pm\pi, \pm 2\pi, \ldots$

This can be written as $\theta = n\pi$, where $n \in \mathbb{Z}$.

Quadrant and angle	Reference angle
1st $0 < \theta < 90°$	θ

2nd $90 < \theta < 180^\circ$	$\pi - \theta$
3rd $180 < \theta < 270^\circ$	$-\pi + \theta \, or \, \pi + \theta$
4th $270 < \theta < 360^\circ$	$2\pi - \theta \, or - \theta$

Procedure to Solve a Trigonometric Equation

1. Solve the equation for the fraction value of the variable.
2. Use a table or calculator to write the exact value of the function.
3. If the angle, as calculated in the 2nd step, is not a quadrantal angle, then find its reference angle.
4. Use the reference angle to find the degree measure of each solution in the interval $0 \le \theta \le 360^\circ$.
5. Add $2n\pi$ to the solution found in steps 2 and 4, to write all possible solutions.

Solution of General Trigonometric Equations

When a trigonometric equation contains more than one trigonometric function, trigonometric identities and algebraic formulae are used to transform such trigonometric equations to an equivalent equation that contains only one trigonometric function.

Extraneous Roots: Sometimes it is necessary to square both sides of a trigonometric equation. In such a case, extraneous roots can occur which are to be discarded. Hence, each value of x must be check by substituting it in the given equation.

For example, $x = 2$ is an equation with a root of 2. On squaring, we get $x^2 = 4$, which gives two roots, 2 and -2. However, the root -2 does not satisfy the equation $x = 2$. Therefore, -2 is an extraneous root.

CHAPTER 10

INVERSE TRIGONOMETRIC FUNCTIONS

INTRODUCTION

In previous chapters, we determined the values of trigonometric functions for given angles. However, in the application of trigonometry, we also need to derive the angles when the value of their trigonometric functions are given. For this purpose, we need to have knowledge of **Inverse trigonometric functions.**

We already know that only one to one functions will have an inverse. If a function is not one to one, it may be possible to restrict its domain to make it one to one, so that its inverse can be found.

In this chapter, we shall define the inverse trigonometric functions.

INVERSE TRIGONOMETRIC FUNCTIONS

The Inverse Sine Function

The graph of $y = sinx$, $-\infty < x < +\infty$, is shown in the Figure 1.

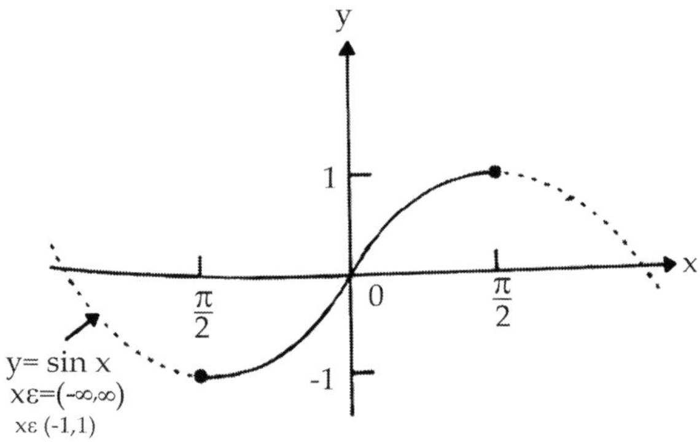

Figure 11.

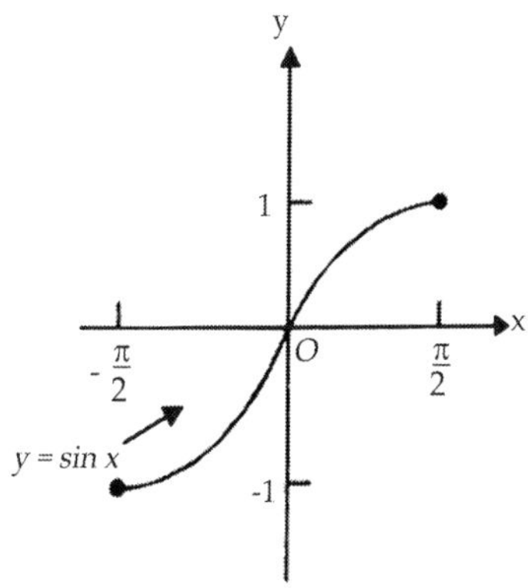

Figure 12.

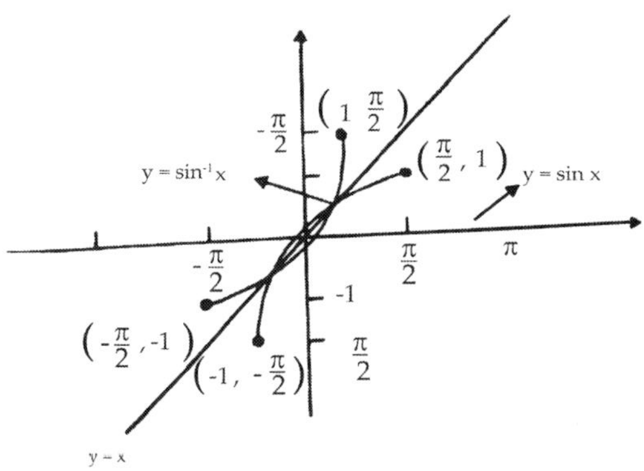

Figure 22.

We observe that every horizontal line between the lines y = 1 and y = -1 intersects the graph infinitely many times. It follows that the sine function is not one-to-one. However, if we restrict the domain of y = sin x to $-\frac{\pi}{2} \le x \le \frac{\pi}{2}$, we get the principal sine function, which is one-to-one, and hence has an inverse.

This inverse function is called the **inverse sine function** and is written as $\sin^{-1} x$ or as sin *x*.

The **inverse sine function** is defined by:

$$y = \sin^{-1} x \text{ if and only if } x = \sin y,$$

where $-\frac{\pi}{2} \le y \le \frac{\pi}{2}$ and $-1 \le x \le 1$.

Here *y* is the angle whose sine is *x*. The domain of the function y = $\sin^{-1} x$ is $-1 \le x \le 1$, and its range is $-\frac{\pi}{2} \le y \le \frac{\pi}{2}$.

The graph of y = $\sin^{-1} x$ is obtained by reflecting the restricted portion of the graph of y = sin x about the line y = x

We notice that the graph of y = sin x is along the x- axis, whereas the graph of y = $\sin^{-1} x$ is along the y- axis.

Note: it must be remembered that $\sin^{-1} \neq (\sin)^{-1}$.

Example 1: Find the value of (i) $\sin^{-1} \frac{\sqrt{3}}{2}$ (ii) $\sin^{-1}\left(-\frac{1}{2}\right)$.

Solution: (i) We want to find the angle y, whose sine is $\frac{\sqrt{3}}{2}$.

$$\Rightarrow \sin y = \frac{\sqrt{3}}{2}, \qquad -\frac{\pi}{2} \le y \le \frac{\pi}{2}$$

$\Rightarrow \qquad y = \frac{\pi}{3}$

$\therefore \sin^{-1}(\frac{\sqrt{3}}{2}) = \frac{\pi}{3}.$

We want to find the angle y whose sine is - $\frac{1}{2}$.

$\Rightarrow \sin y = -\frac{1}{2}, \quad -\frac{\pi}{2} \leq y \leq \frac{\pi}{2}$

$\therefore \qquad y = -\frac{\pi}{6}$

$\therefore \sin^{-1}(-\frac{1}{2}) = -\frac{\pi}{6}.$

The Inverse Cosine Function

The graph of $y = \cos x, -\infty < x < +\infty$, is given below.

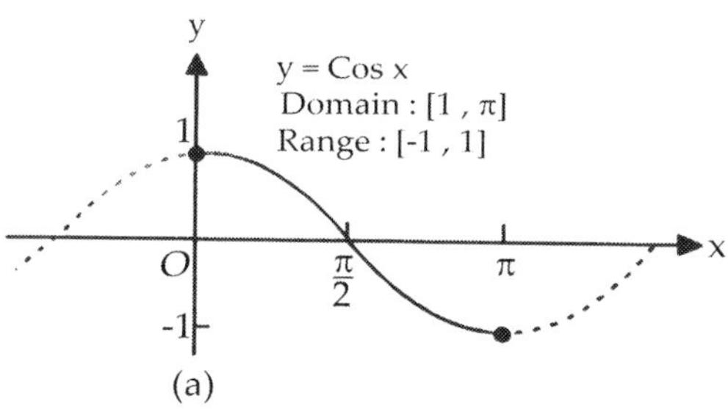

Figure 23.

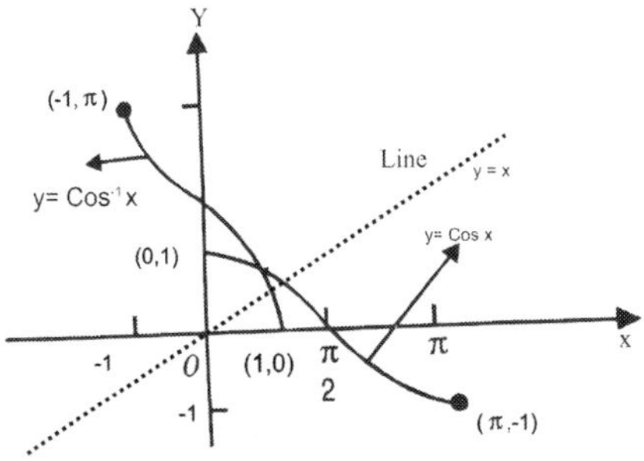

Figure 24.

We observe every horizontal line between the lines y = 1 and y = -1 intersects the graph infinitely many times. It follows that the cosine function is not one-to-one. However, if we restrict the domain of y = cos x to the interval $[0,\pi]$. then the restricted function $y=cos\,x\ ,0\leq x\leq\pi$ is called the **principal cosine function;** which is now one-to-one and hence will have an inverse.

This inverse function is called the inverse cosine function and is written as $\cos^{-1}x$ or $arc\cos x$.

The **Inverse Cosine Function** is defined by:

$y=\cos^{-1}x$, if and only if, $x=\cos y$,

where $0\leq y\leq\pi$ and $-1\leq x\leq 1$.

Here y is the angle whose cosine is x. The domain of function y = $\cos^{-1}x$ is

$-1\leq x\leq 1$ and its range is $0\leq y\leq\pi$.

The graph $y=\cos^{-1}x$ is obtained by reflecting the restricted portion of the graph of $y=$ $cos\,x$ about $y=x$.

We notice that the graph of $y=\cos^{-1}x$ is along the y – axis.

Note: It must be remembered that $\cos^{-1}x\neq(\cos x)^{-1}$.

Example: Find the value of (i) $\cos^{-1}1$ (ii) $\cos^{-1}\left(-\frac{1}{2}\right)$.

Solution: (i) We want to find the angle y whose cosine is 1:

$\Rightarrow \cos y = 1 \qquad\qquad 0 \le y \le \pi$

$\Rightarrow y = 0$

$\therefore \cos^{-1} 1 = 0$.

(ii) We want to find the angle y whose cosine is $-\frac{1}{2}$.

$\Rightarrow \cos y = -\frac{1}{2}$, $0 \le y \le \pi$

$\therefore \quad y = \frac{2\pi}{3}$

$\therefore \quad \cos^{-1}(-\frac{1}{2}) = -\frac{2\pi}{3}$.

Inverse Tangent Function

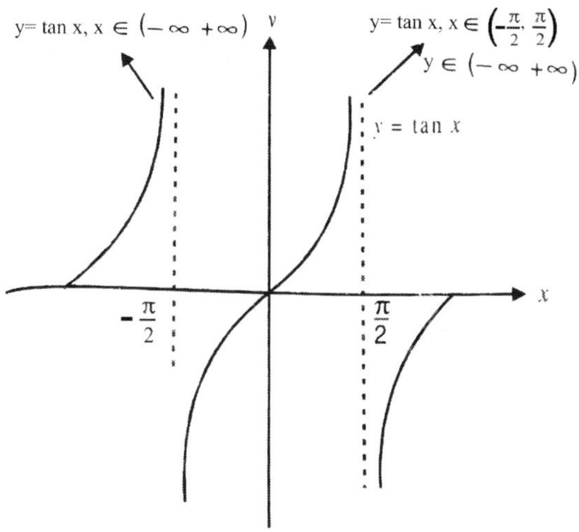

Figure 25.

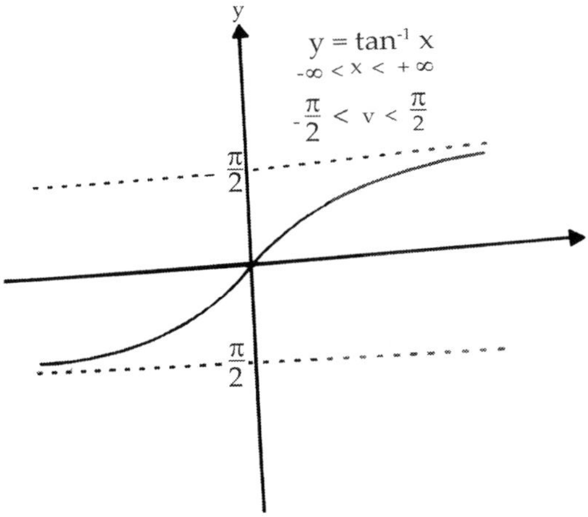

Figure 26.

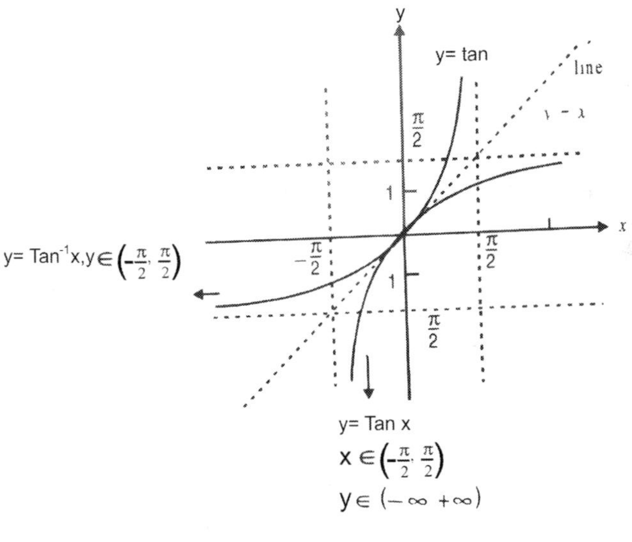

Figure 27.

The graph of $y=\tan x, -\infty<+\infty$, is presented in Figure 25.

Observe that every horizontal line between the lines y = 1 and y = -1 intersects the graph infinitely many times. It follows that the tangent function is not one-to-one. However, if we restrict the domain of y =tan x

to the interval $-\frac{\pi}{2}<x<\frac{\pi}{2}$, the restricted function $y=\tan x,\ \frac{\pi}{2}<x<\frac{\pi}{2}$ is called the **principal tangent function;** which is now one-to-one and hence will have an inverse.

This inverse function is called the **inverse tangent function** and is written as $\tan^{-1} x$ or $arc \tan x$.

The **inverse Tangent function** is defined by:

$y = \tan^{-1} x$ if and only $if\ x = \tan y$,

where $-\frac{\pi}{2} < y < \frac{\pi}{2}$ and $-\infty < x < +\infty$.

Here y is the angle whose tangent is x. The domain of the function y = $\tan^{-1}$ is

$-\infty < x < +\infty$,

and its range is $-\frac{\pi}{2} < y < \frac{\pi}{2}$.

The graph of y = $\tan^{-1} x$ is obtained by reflecting the restricted portion of the graph of y = $\tan x$ about the line y = x.

We notice that the graph of y = $\tan x$ is along the x – axis, whereas the graph of y = $\tan^{-1} x$ is along the y – axis.

Note: it must be remembered that $\tan^{-1} x \neq (\tan x)^{-1}$.

Example: Find the value of (i) $\tan^{-1} 1$ (ii) $\tan^{-1}(-\sqrt{3})$.

Solution: (i) We want to find the angle y, whose tangent is 1.

$\Rightarrow \quad \tan y = 1, \quad -\frac{\pi}{2} < y < \frac{\pi}{2}$

$\Rightarrow \quad y = \frac{\pi}{4}$

$\therefore \quad \tan^{-1} 1 = \frac{\pi}{4}$.

(ii) We want to find the angle y whose tangent is - $\sqrt{3}$.

$\Rightarrow \tan y = -\sqrt{3}$, $\quad -\frac{\pi}{2} < y < \frac{\pi}{2}$.

$\therefore \quad y = \frac{2\pi}{3}$

$\therefore \quad \tan^{-1}\left(-\sqrt{3} = \frac{2\pi}{3}\right)$.

Inverse Cotangent, Secant and Cosecant Functions

These inverse functions are not used frequently, and most calculators do not even have keys to evaluate them. However,, we list their definitions below:

Y =cot x, where $0 < x < \pi$ is called the principal cotangent function,

which is one-to-one and has an inverse.

The inverse cotangent function is defined by:

$y = \cot^{-1} x$, if and only if x = cot y ,

where $0 < y < \pi$ and $-\infty < x < +\infty$.

The student should draw the graph of y = $\cot^{-1} x$ by taking the reflectionof y = cot x on the line y = x. This is left as an exercise for them.

(ii) Inverse Secant Function

$y = sec\, x$, where $0 \leq x \leq \pi$ and x $\neq \frac{\pi}{2}$ is called the **principal secant function,** which is one-to-one and has an inverse.

The inverse secant function is defined by:

$y = \sec^{-1} x$ if and only if $x = sec\, y$,

where $0 \leq y \leq \pi,\ y \neq \frac{\pi}{2}$ and $|x| \leq 1$.

The student should draw the graph of y = $\sec^{-1} x$ by taking the reflection of $y = sec\,x$ on the line of the y = x. This is left as an exercise for them.

(iii) Inverse Cosecant Function

$y = \csc x$, where - $\frac{\pi}{2} \leq y \leq \frac{\pi}{2}$.

$y = \csc^{-1} x$, if and only if $x = \csc y$,

where - $\frac{\pi}{2} \leq y \leq \frac{\pi}{2}$, y ≠ 0 and $|x|\ |\leq|$. The student should draw the graph of y = $\csc^{-1} x$ by taking the reflection of y = csc x on the line y = x. This is left as an exercise for them.

Note. While discussing inverse trigonometric functions, we have seen that there are in general no inverses of trigonometric functions, but restricting their domains to principal functions, allows us to obtain inverse functions.

Domains and Ranges of Principal Trigonometric Functions and Inverse Trigonometric Functions

From the above discussion, we get the following table showing domains and ranges of the principal trigonometric and inverse trigonometric functions.

Table 1

Function	Domain	Range
y = sin x	$\frac{-\pi}{2} \leq x \leq \frac{\pi}{2}$	$-1 \leq x \leq 1$
y = $\sin^{-1} x$	$-1 \leq x \leq 1$	$\frac{-\pi}{2} \leq x \leq \frac{\pi}{2}$
y = cos x	$0 \leq x \leq \pi$	$-1 \leq x \leq 1$

$y = \cos^{-1} x$	$-1 \le x \le 1$	$0 \le x \le \pi$
y = tan x	$\frac{-\pi}{2} < x < \frac{\pi}{2}$	$(-\infty, \infty)$ or $\mathfrak{R}$
$y = \tan^{-1} x$	$(-\infty, \infty)$ or $\mathfrak{R}$	$\frac{-\pi}{2} < x < \frac{\pi}{2}$
y = cot x	$0 < x < \pi$	$(-\infty, \infty)$ or $\mathfrak{R}$
$y = \cot^{-1} x$	$(-\infty, \infty)$ or $\mathfrak{R}$	$0 < x < \pi$
y = sec x	$[0, \eth], x \neq \frac{\pi}{2}$	$y \le -1$ or $y \ge 1$
$y = \sec^{-1} x$	$x \ge -1$ or $x \le 1$	$[0, \pi], x \neq \frac{\pi}{2}$
y = csc x	$-\frac{\pi}{2}, \frac{\pi}{2}$, $x \neq 0$	$y \le -1$ or $y \ge 1$
$y = \csc^{-1} x$	$x \le -1$ or $x \ge 1$	$-\frac{\pi}{2}, \frac{\pi}{2}$, $y \neq 0$

Example 4: Show that $\cos^{-1}\frac{12}{13} = \sin^{-1}\frac{5}{13}$.

Solution: Let $\cos^{-1}\frac{12}{13} = a \Rightarrow \cos a = \frac{12}{13}$

$$\therefore \sin a = \pm\sqrt{1-\cos^2 a} = \pm\sqrt{1-\left(\frac{12}{13}\right)^2}$$

$$= \pm\sqrt{1-\frac{144}{169}}$$

$$= \pm\sqrt{1-\frac{169-144}{169}} = \pm\sqrt{\frac{25}{169}} = \pm\frac{5}{13}$$

$\because$ $\cos a$ is + ve and the domain of a is $[0, \pi]$, in which sine is + ve.

Thus, sin a = $\frac{5}{13}$ $\Rightarrow a = \sin^{-1}\frac{5}{13}$

Hence, $\cos^{-1}\frac{12}{13}=\sin^{-1}\frac{5}{13}$.

Example 5: Find the value of

i) $\sin(\cos^{-1}\frac{\sqrt{3}}{2})$ ii) $\cos(\tan^{-1}0)$ iii) $\sec\left[\sin^{-1}(-\frac{1}{2})\right]$.

Solution:

i) We first find the value of y, whose cosine is $\frac{\sqrt{3}}{2}$.

$\cos y=\frac{\sqrt{3}}{2}$, $\quad 0\leq y\leq\pi$

$\Rightarrow \quad y=\frac{\pi}{6}$

$\Rightarrow (\cos^{-1}\frac{\sqrt{3}}{2} = \frac{\pi}{6}$

$\therefore \quad \sin(\cos^{-1}\frac{\sqrt{3}}{2} = \sin\frac{\pi}{6}=\frac{1}{2}$.

ii) We first find the value of y; whose tangent is 0.

$\tan y = 0$, $-\frac{\pi}{2}<y<\frac{\pi}{2}$

$\Rightarrow \quad y=0$

$\tan^{-1}(0) = 0$

$\cos(\tan^{-1}0) = \cos 0 = 1$.

iii) We first find the value of y, whose sine is $-\frac{1}{2}$.

$\sin y = -\frac{1}{2}$, $\quad -\frac{\pi}{2}\leq y\leq\frac{\pi}{2}$

$$y = -\frac{\pi}{6}$$

$$\sin^{-1}\left(-\frac{1}{2}\right) = -\frac{\pi}{6}$$

$$\text{Sec}\left[\sin^{-1}\left(-\frac{1}{2}\right) = \frac{2}{\sqrt{3}}\right].$$

Example: Prove that the inverse trigonometric functions satisfy the following identities:

i) $\sin^{-1} x = \frac{\pi}{2} - \cos^{-1} x$ and $\cos^{-1} x = \frac{\pi}{2} - \sin^{-1} x$

ii) $\tan^{-1} x = \frac{\pi}{2} - \cot^{-1} x$ and $\cot^{-1} x = \frac{\pi}{2} - \tan^{-1} x$

iii) $\sec^{-1} x = \frac{\pi}{2} - \csc^{-1} x$ and $\csc^{-1} x = \frac{\pi}{2} - \sec^{-1} x$.

Proof:

Consider the right triangle given in figure. Angles α and β are acute and complementary.

$$\Rightarrow \quad \alpha + \beta = \frac{\pi}{2}$$

$$\Rightarrow \quad \alpha + \beta = \frac{\pi}{2} - \beta \quad \text{and} \quad \beta = \frac{\pi}{2} - \alpha.$$

Now, $\sin \alpha = \sin\left(\frac{\pi}{2} - \beta\right) = \cos \beta = x$ (say)

$\alpha = \sin^{-1} x$ and $\beta = \cos^{-1} x$.

Thus, from (i) we have:

$\sin^{-1} x = \frac{\pi}{2} - \cos^{-1} x$ and $\cos^{-1} x = \frac{\pi}{2} - \sin^{-1} x$.

In a similarly way, we can derive identities (ii) and (iii).

ADDITION AND SUBTRACTION FORMULAS

1) Prove that $\sin^{1} A+\sin^{-1} B=\sin^{-1}(A\sqrt{1-B^2}+B\sqrt{1-A^2})$.

Proof: Let $\sin^{-1} A = x \quad \Rightarrow \quad sin\ x = A$

and $\sin^{-1} B = y \quad \Rightarrow \quad sin\ y = B$.

Now $\cos x = \pm\sqrt{1-\sin^2 x} = \sqrt{1-A^2}$

For sin x = A, the domain is $\left[-\frac{\pi}{2},\frac{\pi}{2}\right]$, in which cosine is + ve,

$\therefore \quad \cos x = \sqrt{1-A^2}$

Similarly, cos y = $\sqrt{1-B^2}$.

Now sin (x + y) = sin x cos y + cos x sin y

A = $\sqrt{1-B^2}$ + B $\sqrt{1-A^2}$

$\Rightarrow \quad x+y = \sin^{-1}(A\sqrt{1-B^2} + B\sqrt{1-A^2}) \quad \therefore$

$$\sin^{-1} A+\sin^{-1} B=\sin^{-1}(A\sqrt{1-B^2}+B\sqrt{1-A^2})$$

In a similar way, we can prove that

$$\sin^{-1} A-\sin^{-1} B=\sin^{-1}(A\sqrt{1-B^2}-B\sqrt{1-A^2})$$

$$\cos^{-1} A+\cos^{-1} B=\cos^{-1}(AB\sqrt{(1-A^2)\,(1-B^{2)}}\ )$$

$$\cos^{-1} A-\cos^{-1} B=\cos^{-1}(AB\sqrt{(1-A^2)\,(1-B^{2)}}\ ).$$

5) Prove that:

$$\tan^{-1} A+\tan^{-1} B=\tan^{-1}\frac{A+B}{1-AB}.$$

Proof: Let $\tan^{-1} A = x \quad \Rightarrow \quad$ tan x = A

and $\tan^{-1} B = y \quad \Rightarrow \quad$ tan y = B.

Now, tan(x+y) = 1 $\dfrac{\tan x + \tan y}{1 - \tan x \tan y} = \dfrac{A+B}{1-AB}$

$$\Rightarrow \qquad x + y = \tan^{-1}\frac{A+B}{1-AB}.$$

In a similar way, we can prove that:

6) $\tan^{-1} A - \tan^{-1} B = \tan^{-1}\dfrac{A-B}{1+AB}$.

Cor. Putting A= B in

$\tan^{-1} A + \tan^{-1} B = \tan^{-1}\dfrac{A+B}{1-AB}$, we get

$$\tan^{-1} A + \tan^{-1} A = \tan^{-1}\frac{A+A}{1-A^2}$$

$$\Rightarrow \quad 2\tan^{-1} A = \tan^{-1}\frac{2A}{1-A^2}.$$

SUMMARY

After completing this chapter, students will be able to:

- Understand inverse functions.
- Construct graphs of inverse trigonometric functions.
- Apply sum and difference formulas to inverse problems.

Key Terms and Formulas

Function	**Domain**	**Range**
$y = \sin^{-1} x$	$-1 \le x \le 1$	$\dfrac{-\pi}{2} \le x \le \dfrac{\pi}{2}$
$y = \cos^{-1} x$	$-1 \le x \le 1$	$0 \le x \le \pi$
$y = \tan^{-1} x$	$(-\infty, \infty)$ or $\Re$	$\dfrac{-\pi}{2} < x < \dfrac{\pi}{2}$

$y = \cot^{-1} x$	$(-\infty,\infty)$ or $\Re$	$0 < x < \pi$
$y = \sec^{-1} x$	$x \geq -1$ or $x \leq 1$	$[0,\pi], x \neq \frac{\pi}{2}$
$y = \csc^{-1} x$	$x \leq -1$ or $x \geq 1$	$-\frac{\pi}{2}, \frac{\pi}{2}$, $y \neq 0$

$$\sin^{-1} A + \sin^{-1} B = \sin^{-1}(A\sqrt{1-B^2} + B\sqrt{1-A^2})$$

$$\sin^{-1} A - \sin^{-1} B = \sin^{-1}(A\sqrt{1-B^2} - B\sqrt{1-A^2})$$

$$\cos^{-1} A + \cos^{-1} B = \cos^{-1}(AB\sqrt{(1-A^2)\,(1-B^{2)}}\,)$$

$$\cos^{-1} A - \cos^{-1} B = \cos^{-1}(AB\sqrt{(1-A^2)\,(1-B^{2)}}\,)$$

$$\tan^{-1} A + \tan^{-1} B = \tan^{-1}\frac{A+B}{1-AB}$$

$$\tan^{-1} A - \tan^{-1} B = \tan^{-1}\frac{A-B}{1+AB}$$

i) $\sin^{-1} x = \frac{\pi}{2} - \cos^{-1} x$ and $\cos^{-1} x = \frac{\pi}{2} - \sin^{-1} x$

ii) $\tan^{-1} x = \frac{\pi}{2} - \cot^{-1} x$ and $\cot^{-1} x = \frac{\pi}{2} - \tan^{-1} x$

iii) $\sec^{-1} x = \frac{\pi}{2} - \csc^{-1} x$ and $\csc^{-1} x = \frac{\pi}{2} - \sec^{-1} x.$

CHAPTER 11

SPHERICAL TRIGONOMETRY AND APPLICATIONS OF TRIGONOMETRY

SPHERICAL TRIGONOMETRY

Spherical trigonometry is the branch of spherical geometry that deals with the relationships between trigonometric functions of the sides and angles of the spherical polygons (particularly spherical triangles) defined by a number of intersecting great circles on the sphere. Spherical trigonometry is of great importance for calculations in astronomy, geodesy and navigation.

The origins of spherical trigonometry in Greek mathematics, and the major developments in Islamic mathematics, are discussed fully in the History of trigonometry and Mathematics in medieval Islam. The subject came to fruition in early modern times with important developments by John Napier, Delambre and others, and attained an essentially complete form by the end of the nineteenth century with the publication of Todhunter's textbook on Spherical trigonometry for the use of colleges and schools. This book is now readily available on the web. The only significant developments since then have been the application of vector methods for the derivation of theorems and the use of computers to carry through lengthy calculations.

Spherical angle

A Spherical angle is the angle between any two intersecting great circles.

Two planes α and β with a common line of intersection AB are said to form a dihedral angle, AB is called the edge of the dihedral angle, and the planes forming the dihedral angle are called its faces.

The plane angle of the dihedral angle is the angle formed by two rays, one on each face, perpendicular to the edge at the same point. The dihedral angle is equal its plane angle.

A trihedral angle is formed by three planes which intersect so as to have only one point is common. The point forming a trihedral angle is known as the vertex of the trihedral angles.

A spherical triangle is that part of the surface of a sphere which is bounded by the arcs of three great circles. Note that a spherical triangle is the intersection of the surface of the sphere with the trihedral angle whose vertex is *O*, the center of the sphere, and whose faces are the planes of the great circles.

The side of a spherical triangle is equal to the angle subtended by that side at the center of the sphere. By definition, a spherical angle is the angle between any two intersecting great circles. It can be measured by measuring angle between the tangents at the point of intersection of the great circles, or by measuring the dihedral angle of the trihedral angle, whose vertex is at the center of the sphere.

Spherical polygons

A spherical polygon is a polygon on the surface of the sphere, defined by a number of great-circle arcs that are the intersection of the surface with planes through the center of the sphere. Such polygons may have any number of sides. Two planes define a lune, also called a "digon" or bi-angle, the two-sided analogue of the triangle: a familiar example is the curved surface of a segment of an orange. Three planes define a spherical triangle, the principal subject here. Four planes define a spherical quadrilateral: such a figure, and higher sided polygons, can always be treated as a number of spherical triangles.

From this point the chapter will be restricted to spherical triangles, denoted simply as triangles.

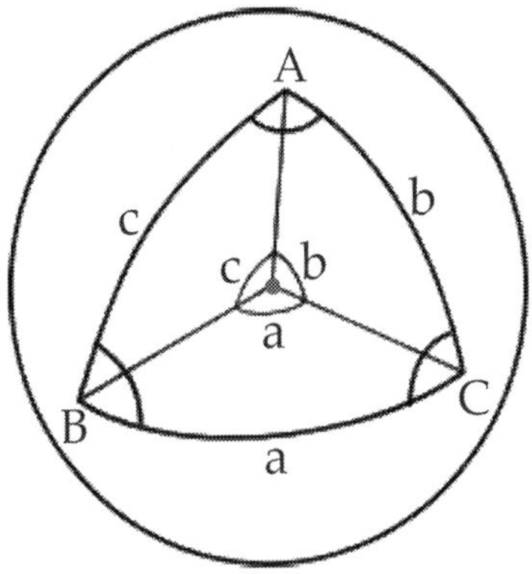

Figure 1

The basic triangle on a unit sphere:

- Both vertices and angles at the vertices are denoted by the same upper-case letters A, B and C.
- The angles A, B, C of the triangle are equal to the angles between the planes that intersect the surface of the sphere, or, equivalently, the angles between the tangent vectors of the great circle arcs where they meet at the vertices. Angles are in radians. The angles of proper spherical triangles are (by convention) less than π so that $\pi < A+B+C < 3\pi$.
- The sides are denoted by lower-case letters a,b,c. Their lengths are numerically equal to the radian measure of the angles that the great circle arcs subtend at the center. The sides of proper spherical triangles are (by convention) less than π so that $0 < a+b+c < 3\pi$.
- The radius of the sphere is taken as unity. For specific practical problems on a sphere of radius R, the measured lengths of the sides must be divided by R before using the identities given below. Likewise, after a calculation on the unit sphere, the sides a, b, c must be multiplied by R.

Polar triangles

The polar triangle associated with a triangle ABC is defined as follows. Consider the great circle that contains the side BC. This great circle is defined by the intersection of a diametral plane with the surface. Draw the normal to that plane at the center: it intersects the surface at two points and the point that is on the same side of the plane as A is (conventionally) termed the pole of A, and it is denoted by A. The points B and C are similarly defined.

The triangle *ABC* is the polar triangle corresponding to triangle ABC. A very important theorem proves that the angles and sides of the polar triangle are given by:

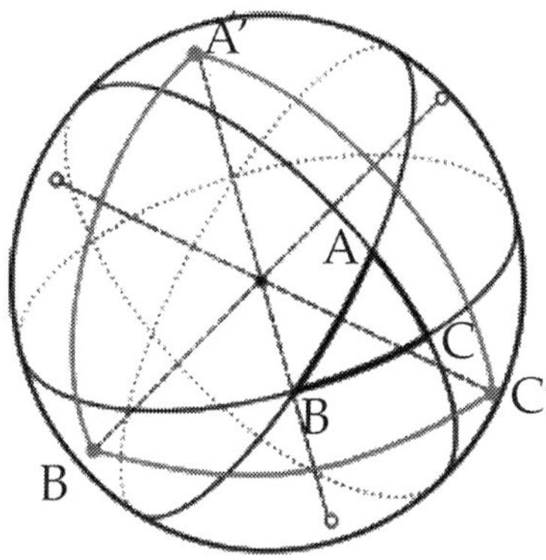

Figure 2

$A' = \pi - a$, $B' = \pi - b$, $C' = \pi - c$, $a' = \pi - A$, $b' = \pi - B$, $c' = \pi - C$.

Therefore, if any identity is proved for the triangle ABC, then we can immediately derive a second identity by applying the first identity to the polar triangle by making the above substitutions. This is how the supplemental cosine equations are derived from the cosine equations. Similarly, the identities for a quadrantal triangle can be derived from those for a right-angled triangle. The polar triangle of a polar triangle is the original triangle.

The cosine formula

The cosine rule is the fundamental identity of spherical trigonometry: all other identities, including the sine rule, may be derived from the cosine rule.

There are many ways of deriving the fundamental cosine and sine rules and the other rules developed in the following sections. For example, To dhunter gives two proofs of the cosine rule and two proofs of the sine rule. The page on the spherical law of cosines gives four different proofs of the cosine rule. Text books on geodesy and spherical astronomy give different proofs and online resources of Math World provide yet more. There are even more exotic derivations, such as that of Banerjee, who derives the formulae using the linear algebra of projection matrices and also quotes methods in differential geometry and the group theory of rotations.

Proof: Suppose we have a spherical triangle ABC on a sphere of unit radius. Let the measure of the sides BC, CA and AB be a, b and c respectively. Let the measure of the dihedral angle defined by the planes OAB and OAC be donated by a.

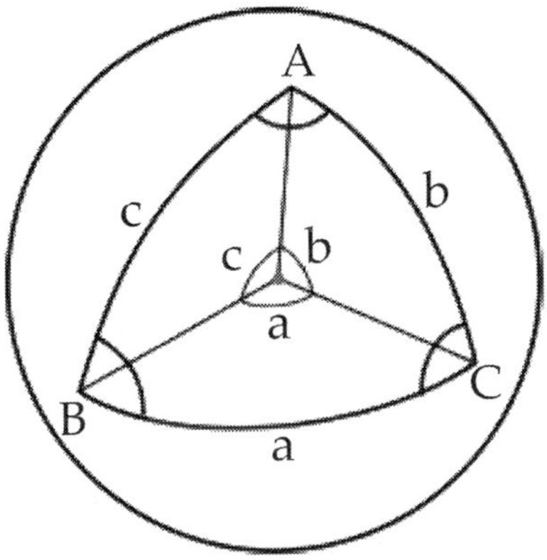

Figure 3

Similarly, we define β and γ. Let $\overline{OA} = \mathbf{I}, \overline{OB}$ **= m and** $\overline{OC}$ **= n.** As the sphere is of unit radius. **I, m, n** are unit vectors.

Now, from vector analysis we get:

(I x m). (I x n) = $\begin{vmatrix} I.I & I.n \\ m.I & m.n \end{vmatrix} = \begin{vmatrix} 1 & \cos b \\ \cos c & \cos a \end{vmatrix}$

$= \cos a - \cos b \cos c$ (1)

Also, **I x m** is a vector perpendicular to the plane AOB (drawn inward) and its magnitude is sin c. Similarly, **l x n** is a vector perpendicular to the plane AOC (drawn outward) and its magnitude is sin b.

Therefore, **(l x m). (l x n)** = sin b sin c cos A. (2)

From (1) and (2), we obtain

$\cos a - \cos b \cos c = \sin b \sin c \cos A$.

Thus, $\cos a = \cos b \cos c + \sin b \sin c \cos A$. (3)

In a similar manner, we can prove that

$\cos b = \cos c \cos a + \sin c \sin a \cos B$ (4)

$\cos c = \cos a \cos b + \sin a \sin b \cos C$. (5)

Results (3), (4) and (5) are known as the cosine formulas.

The sine formula

Consider Figure 3 again, we will obtain

Now, **(I x m) x (I x n)** $= [\mathbf{I\,m\,n}]\mathbf{I} - [\mathbf{I\,m\,I}]\mathbf{n}$

$= [\mathbf{I\,m\,n}]$ **I.**

The left-hand side is perpendicular to the normals to the planes *AOB* and *AOC* and is, therefore, in the direction of **I,** while the magnitude of the vector is sin c and b sin A.

Thus, $\sin c \sin b \sin A\, \boldsymbol{I} = [\boldsymbol{l\ m\ n}]\boldsymbol{I}$

$\sin b \sin c \sin A = [\mathbf{l\,m\,n}]$. (1)

Similarly, $\sin c \sin a \sin B = [\mathbf{l\,m\,n}]$ (2)

From (1) and (2), $sin\, b\, sin\, c\, sin\, A = sin\, c\, sin\, a\, sin\, b$

Therefore, $\frac{\sin A}{sin\, a} = \frac{\sin B}{sin\, b}$

Similarly, $\frac{\sin B}{sin\, b} = \frac{\sin C}{sin\, c}$.

Hence $\frac{\sin A}{sin\, a} = \frac{\sin B}{sin\, b} = \frac{\sin C}{sin\, c}$,
which is known as the sine formula.

The four-parts (or the cotangent) formula

From (5) and (3), we have

$\cos c = \cos a \cos b + \sin a \sin b \cos C$

$\cos a = \cos b \cos c + \sin b \sin c \cos A$.

Eliminating cos c from the above equations, we get:

Cos $a = \cos b(\cos a \cos b + \sin a \sin b \cos C) + \sin b \sin c \cos A$.

Thus, cos a $(1 - \cos^2 b = sina \sin b \cos b \cos C + \sin b \sin c \cos A$.

Or $\cos a \sin^2 b = \sin a \sin b \cos b \cos C + \sin b \sin c \cos A$.

Dividing both sides by sin a $\sin b$, we get

$\cot a \sin b = \cos b \cos C + \frac{sinc}{sina} \cos A$.

Or $\cot a \sin b$ = cos b cosC + $\frac{sinC}{sinA}$ Cos A (using the sine formula)

Or $\cos b \cos C = \sin b \cot a - \sin C \cot A$.

This is known as the four-part (or the **cotangent**) formula.

Latitude and longitude

For many practical problems, the Earth is regarded as a sphere spinning about a dinner PQ, where P is called the North Pole and Q the South Pole. The great circle, whose plane is perpendicular to PQ, is called the equator. Any semi-great circle with diameter PQ is called a meridian. In particular, the meridian that passes through the fundamental instrument (the transit circle) of the Greenwich Observatory is, by universal agreement, recorded as the principal or standard meridian. Let PGKQ intersect the equator in K, let PHQL be any other meridian cutting the equator at L. The measure of the angle KOL is defined to be the longitude of the meridian PHQ and it can be described equally well as the length of the equatorial arc KL or the magnitude of the spherical angle KPL. Longitudes are measured from 0 to 180 east (E) of the Greenwich Meridian and from 0 to 180 west (W). Thus, from the figure, the longitude of the meridian PHQ is about 100 E and that the meridian PMQ is about 60 W. All planes on the same meridian have the same longitude and the meridian on which a particular place is situated is specified with reference to the principal meridian PGQ. To specify completely the position of a place on the surface of the Earth, we are required to describe its position on the meridian of the longitude. This is done with reference to the equator. Consider a location J on the meridian PHQ. The meridian through J cuts the equator in L and the measure of the angle LOJ, or the length of the great circle arc LJ, is called the latitude (lat) of J.

If J is between the equator of the North Pole P, the latitude is said to be north (N). A location such as R, between the equator and the South Pole Q, is said to be in south (S). In this way, the position of any point on the surface of the Earth is referred to using great circles, the equator and the meridian of Greenwich.

Let ∅ denote the latitude of *J*, then the measure of angle LOJ or LJ is ∅. Since OP is perpendicular to the plane of the equator, $m< POL = 90°$ and therefore, m<POJ or the spherical and PJ is the **Co-latitude (Colat)** of J. We thus have:

Colat = 90^0 – lat.

All places that have the same latitude lie on a small circle parallel to the equator, known as the **parallel of latitude.** Thus, all places with the same latitude as Greenwich lie on *MGHX*. If ∅ denotes the latitude of Greenwich, then the length of the corresponding equatorial arc *LY* is given by:

$HX = LY \cos \varnothing.$

On the great circle, the distance between two points subtending an angle of one minute of arc at the center of the Earth is taken as the unit distance on the surface. Calculate the distance between *H* and *X* expressed in nautical miles (or minutes of arc) and measured along the parallel of latitude.

Some related proofs:

Prove that in a spherical triangle ABC:

$$\sin\frac{A}{2} = \sqrt{\frac{\sin(s-b)\sin(s-c)}{\sin b \sin c}}$$

$$\cos\frac{A}{2} = \sqrt{\frac{\sin s \sin(s-a)}{\sin b \sin c}}$$

a) $\tan\frac{A}{2} = \sqrt{\frac{\sin(s-b)\sin(s-c)}{\sin s \sin(s-c)}}$, where $2s = a+b+c$.

Proof:

Consider a) $\sin\frac{A}{2} = \sqrt{\frac{\sin(s-b)\sin(s-c)}{\sin b \sin c}}$.

We know that $2\sin^2\frac{A}{2} = 1 - \cos A$

$$= 1 - \frac{\cos a - \cos b \cos c}{\sin b \sin c}$$

$$= \frac{\sin b \sin c - \cos a + \cos b \cos c}{\sin b \sin c}$$

$$=\frac{\cos b\,\cos c+\sin b\sin c-\cos a}{\sin b\sin c}$$

$$=\frac{\cos(b-c)-\cos a}{\sin b\sin c}$$

$$=\frac{2\sin\dfrac{a+b-c}{2}\sin\dfrac{a-b+c}{2}}{\sin b\sin c}$$

$$=\frac{2\sin(s-c)\sin(s-b)}{\sin b\sin c},$$

$$\text{or }\sin^2\frac{A}{2}=\frac{\sin(s-b)\sin(s-c)}{\sin b\sin c}$$

$$\sin\frac{A}{2}=\sqrt{\frac{\sin(s-b)\sin(s-c)}{\sin b\sin c}}.$$

Oblique triangles

The solution of triangles is the principal purpose of spherical trigonometry: given three, four or five elements of the triangle, determine the remainder. The case of five given elements is trivial, requiring only a single application of the sine rule. For four given elements, there is one non-trivial case, which is discussed below. For three given elements, there are six cases: three sides, two sides and an included or opposite angle, two angles and an included or opposite side, or three angles (the last case has no analogue in planar trigonometry). No single method solves all cases. The figure below shows the seven non-trivial cases: in each case the given sides are marked with a cross-bar and the given angles with an arc (the given elements are also listed below the triangle). There is a full discussion of the solution of oblique triangles in Todhunter.

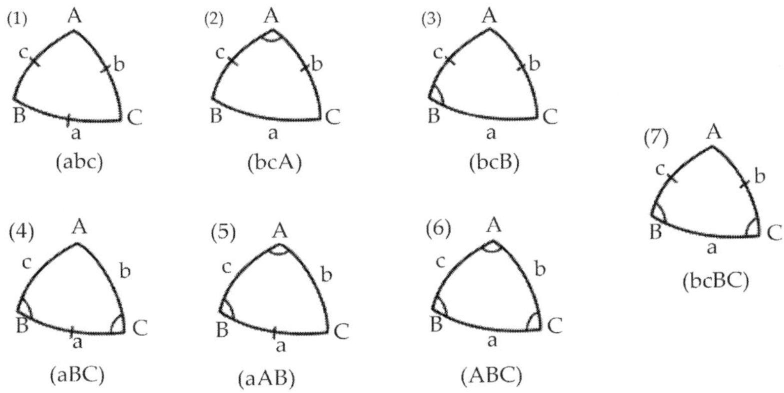

Figure 4

- Case 1: Three sides given. The cosine rule may be used to give the angles A, B, and C but, to avoid ambiguities, the half angle formulae are preferred.
- Case 2: Two sides and an included angle given. The cosine rule gives a and then we are back to Case 1.
- Case 3: Two sides and an opposite angle given. The sine rule gives C and then we have Case 7. There are either one or two solutions.
- Case 4: Two angles and an included side given. The four-part cotangent formulae for sets (cBaC) and (BaCb) give c and b, then A follows from the sine rule.
- Case 5: Two angles and an opposite side given. The sine rule gives b and then we have Case 7 (rotated). There are either one or two solutions.
- Case 6: Three angles given. The supplemental cosine rule may be used to give the sides a, b, and c but, to avoid ambiguities, the half-side formulae are preferred.
- Case 7: Two angles and two sides given. Use Napier's analogies for a and A.

The solution methods listed here are not the only possible choices: many others are possible. In general, it is better to choose methods that avoid taking an inverse sine because of the possible ambiguity between an angle and its supplement. The use of the half-angle formulae is often advisable because half-angles will be less than $\pi/2$ and therefore free from ambiguity. There is a full discussion in Todhunter relating to this.

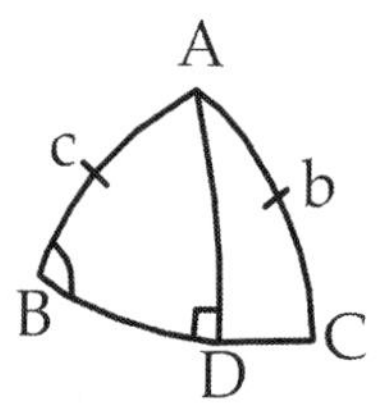

Figure 5

Solution by right-angled triangles

Another approach is to split the triangle into two right-angled triangles. For example, take Case 3 where b, c, B are given. Construct the great circle from A that is normal to the side BC at the point D. Use Napier's rules to solve the triangle ABD: use c and B to find the sides AD, BD and the angle BAD. Use Napier's rules to then solve the triangle ACD: that is, use AD and b to find the side DC and the angles C and DAC. The angle A and side a follow this.

Study of spherical triangles

Spherical trigonometry involves the study of spherical triangles, which are formed by the intersection of three great circle arcs on the surface of a sphere. Spherical triangles were subject to intense study from antiquity because of their usefulness in navigation, cartography, and astronomy.

The angles of a spherical triangle are defined by the angle of intersection of the corresponding tangent lines to each vertex. The sum of the angles of a spherical triangle is always greater than the sum of the angles in a planar triangle (π radians, equivalent to two right angles). The amount by which each spherical triangle exceeds two right angles (in radians) is known as its spherical excess. The area of a spherical triangle is given by the product of its spherical excess E and the square of the radius r of the sphere it resides on—in symbols, Er^2.

By connecting the vertices of a spherical triangle with the center O of the sphere that it resides on, a special "angle" known as a trihedral angle is formed. The central angles (also known as dihedral angles) between each pair of line segments OA, OB, and OC are labeled α, β, and γ to correspond to the sides (arcs) of the spherical triangle labeled a, b, and c, respectively. As a trigonometric function of a central angle and its corresponding arc have the same value, spherical trigonometry formulas are given in terms of the spherical angles A, B, and C and, interchangeably, in terms of the arcs a, b, and c and the dihedral angles α, β, and γ. Furthermore, most formulas from plane trigonometry have an analogous representation in spherical trigonometry. For example, there is a spherical law of sines and a spherical law of cosines.

Common spherical trigonometry formulas

Law of sines $\frac{\sin a}{\sin A} = \frac{\sin b}{\sin B} = \frac{\sin c}{\sin C}$

Law of cosine $\cos a = \cos b \cos c + \sin b \sin c \cos A$

$$\cos b = \cos a \cos c + \sin a \sin c \cos B$$

$$\cos c = \cos a \cos b + \sin a \sin b \cos C$$

Half angle formulas $\tan\left(\frac{A}{2}\right) = \sqrt{\frac{\sin(s-b)\sin(s-c)}{\sin s \sin(s-a)}}$

$$\tan\left(\frac{B}{2}\right) = \sqrt{\frac{\sin(s-a)\sin(s-c)}{\sin s \sin(s-b)}}, \quad \text{where } s = \frac{a+b+c}{2}$$

$$\tan\left(\frac{C}{2}\right) = \sqrt{\frac{\sin(s-a)\sin(s-b)}{\sin s \sin(s-c)}}$$

Half side formulas

$$\tan\left(\frac{a}{2}\right) = \sqrt{\frac{-\cos S \cos(S-A)}{\cos(S-B)\cos(S-C)}}$$

$$\tan\left(\frac{b}{2}\right)=\sqrt{\frac{-\cos S\cos(S-B)}{\cos(S-A)\cos(S-C)}}\text{, where } S=\frac{A+B+C}{2}$$

$$\tan\left(\frac{c}{2}\right)=\sqrt{\frac{-\cos S\cos(S-C)}{\cos(S-A)\cos(S-B)}}$$

As was described for a plane triangle, the known values involving a spherical triangle are substituted into the analogous spherical trigonometry formulas, such as the laws of sines and cosines, and the resulting equations are then solved for the unknown quantities.

Many other relations exist between the sides and angles of a spherical triangle. It is worth mentioning the Napier's analogies (derivable from the spherical trigonometry half-angle or half-side formulas), which are particularly well suited for use with logarithmic tables.

Napier's analogies

$$\tan\left(\frac{a}{2}\right)\cos\left(\frac{B-C}{2}\right)=\tan\left(\frac{b+c}{2}\right)\cos\left(\frac{B+C}{2}\right)$$

$$\tan\left(\frac{a}{2}\right)\sin\left(\frac{B-C}{2}\right)=\tan\left(\frac{b-c}{2}\right)\sin\left(\frac{B+C}{2}\right)$$

$$\cot\left(\frac{A}{2}\right)\cos\left(\frac{b-c}{2}\right)=\tan\left(\frac{B+C}{2}\right)\cos\left(\frac{b+c}{2}\right)$$

$$\cot\left(\frac{A}{2}\right)\sin\left(\frac{b-c}{2}\right)=\tan\left(\frac{B-C}{2}\right)\sin\left(\frac{b+c}{2}\right)$$

Applications of trigonometry

Trigonometry is a branch of mathematics that studies the interaction between the sides and the angles in triangles. Trigonometry defines the trigonometric functions, which express these associations and have applicability to cyclical phenomena, such as waves.

Among the lay community of non-mathematicians and non-scientists, trigonometry is known mainly for its application to measurement problems, yet it is also often used in ways that are far more delicate, such as its place in the theory of music; and other more technical uses, such as number theory. The mathematical topics of Fourier series and Fourier transforms rely a lot on knowledge of trigonometric functions and are applied in a number of areas, including statistics. Scientific fields that make use of trigonometry include:

- Acoustics
- Architecture
- Astronomy
- Cartography
- Civil engineering
- Geophysics
- Crystallography
- Electrical engineering
- Electronics
- Land surveying and geodesy
- Many physical sciences
- Mechanical engineering
- Machining
- Medical imaging
- Number theory
- Oceanography
- Optics
- Pharmacology
- Probability theory
- Seismology
- Statistics

- Visual perception.

The fact that these fields involve trigonometry does not mean knowledge of trigonometry is needed in order to learn anything about them. It does mean that various aspects in these fields cannot be understood without trigonometry. For example, a professor of music may perhaps know nothing of mathematics, but would probably know that Pythagoras was the earliest known contributor to the mathematical theory of music.

In some of the fields listed above, it is easy to imagine how trigonometry can be used. For example, in navigation and land surveying, the occasions for the use of trigonometry are in at least some cases simple enough that they can be described in a beginners level trigonometry textbook. In the case of music theory, the application of trigonometry is related to the work begun by Pythagoras, who observed that the sounds made by plucking two strings of different lengths are consonant if both lengths are small integer multiples of a common length. The resemblance between the shape of a vibrating string and the graph of the sine function is no mere coincidence. In oceanography, the resemblance between the shapes of some waves and the graph of the sine function is also not coincidental. In some other fields, among them climatology, biology, and economics, seasonal periodicities are studied. The study of these often involves the periodic nature of the sine and cosine functions.

Fourier series

Many fields make use of trigonometry in more advanced ways than can be discussed here. Often those involve what are called Fourier series, after the 18th- and 19th-century French mathematician and physicist Joseph Fourier. Fourier series have a surprisingly diverse array of applications in many scientific fields, in particular, in all of the phenomena involving seasonal periodicities mentioned above. They are also useful in wave motion, and hence in the study of radiation, acoustics, seismology, modulation of radio waves in electronics, and electric power engineering.

Fourier series are also applicable to subjects whose connection with wave motion is far from obvious. One ubiquitous example is digital compression, whereby images, audio and video data are compressed into a much smaller size, which makes their transmission feasible over telephone, internet and broadcast networks. Another example, mentioned above, is diffusion. Among others are: the geometry of numbers, isoperimetric problems, recurrence of random walks, quadratic reciprocity, the central limit theorem and Heisenberg's inequality.

Fourier transforms

A more abstract concept than the Fourier series is the idea of Fourier transform. Fourier transforms involve integrals rather than sums, and are used in a similarly diverse array of scientific fields. Many natural laws are expressed by relating rates of change of quantities to the quantities themselves. For example, the rate of change of the population is sometimes jointly proportional to (1) the present population and (2) the amount by which the present population falls short of the carrying capacity. This kind of relationship is called a differential equation. If, given this information, one tries to express population as a function of time, one is trying to "solve" the differential equation. Fourier transforms may be used to convert some differential equations to algebraic equations, to be solved by known methods. Fourier transforms have many uses in almost any scientific context in which the words spectrum, harmonic, or resonance are encountered. There is an enormous number of uses of trigonometry and trigonometric functions. For instance, the technique of triangulation is used in astronomy to measure the distance to nearby stars, in geography to measure distances between landmarks, and in satellite navigation systems. The sine and cosine functions are fundamental to the theory of periodic functions, such as those that describe sound and light waves.

Fields that use trigonometry or trigonometric functions include astronomy (particularly for locating apparent positions of celestial objects, in which spherical trigonometry is essential) and hence navigation (on the oceans, in aircrafts, and in space), music theory, audio synthesis, acoustics, optics, electronics, biology, medical imaging (CAT scans and ultrasound), pharmacy, chemistry, number theory (and hence cryptology), seismology, meteorology, oceanography, many physical sciences, land surveying and geodesy, architecture, image compression, phonetics, economics, electrical engineering, mechanical engineering, civil engineering, computer graphics, cartography, crystallography and game development.

Statistics, including mathematical psychology

Intelligence quotients are sometimes held to be distributed according to the bell-shaped curve. About 40% of the area under the curve is in the interval from 100 to 120; correspondingly, about 40% of the population scores between 100 and 120 on IQ tests. Nearly 9% of the area under the curve is in the interval from 120 to 140; correspondingly, about 9% of the population scores between 120 and 140 on IQ tests, etc. Similarly, many other things are distributed according to the "bell-shaped curve", including measurement errors in many physical measurements. Why the ubiquity of the "bell-shaped curve"? There is a theoretical reason for this, and it involves Fourier transforms and hence trigonometric functions. That is one of a variety of applications of Fourier transforms in statistics. Trigonometric functions are also applied when statisticians study seasonal periodicities, which are often represented by Fourier series.

Number theory

There is a hint of a connection between trigonometry and number theory. Loosely speaking, one could say that number theory deals with qualitative properties rather than quantitative properties of numbers. Discard the ones that are not in lowest terms; keep only those that are in lowest terms:

Then bring in trigonometry:

The value of the sum is −1, because 42 has an odd number of prime factors and none of them is repeated: $42 = 2 \times 3 \times 7$. If there had been an even number of non-repeated factors then the sum would have been 1; if there had been any repeated prime factors (e.g., $60 = 2 \times 2 \times 3 \times 5$), then the sum would have been 0; the sum is the Möbius function evaluated at 42. This hints at the possibility of applying Fourier analysis to number theory.

Solving non-trigonometric equations

Various types of equations can be solved using trigonometry.

For example, a linear difference equation or linear differential equation with constant coefficients has solutions expressed in terms of the eigenvalues of its characteristic equation; if some of the eigenvalues are complex, the complex terms can be replaced by trigonometric functions of real terms, showing that the dynamic variable exhibits oscillations. Similarly, cubic equations with three real solutions have an algebraic solution that is unhelpful in that it contains cube roots of complex numbers; again, an alternative solution exists in terms of trigonometric functions of real terms.

Acoustics

Acoustics is the interdisciplinary science that deals with the study of all mechanical waves in gases, liquids, and solids, including topics such as vibration, sound, ultrasound and infrasound. A scientist who works in the field of acoustics is an acoustician, while someone working in the field of acoustics technology may be called an acoustical engineer. The application of acoustics is present in almost all aspects of modern society, with the most obvious being the audio and noise control industries.

Hearing is one of the most crucial means of survival in the animal world, and speech is one of the most distinctive characteristics of human development and culture. Accordingly, the science of acoustics spreads across many facets of human society—music, medicine, architecture, industrial production, warfare and more. Likewise, animal species, such as songbirds and frogs, use sound and hearing as a key element of mating rituals or marking territories. Art, craft, science and technology have provoked one another to advance as a whole, as in many other fields of knowledge. Robert Bruce Lindsay's 'Wheel of Acoustics' is a well-accepted overview of the various fields in acoustics.

The word "acoustic" is derived from the Greek word ἀκουστικός (akoustikos), meaning "of or for hearing, ready to hear and that from ἀκουστός (akoustos), "heard, audible", which in turn derives from the verb ἀκούω (akouo), "I hear". The Latin synonym is "sonic", after which the term sonics used to be a synonym for acoustics, and later, a branch of acoustics. Frequencies above and below the audible range are called "ultrasonic" and "infrasonic", respectively.

Architecture

Latin architectura, from the Greek ἀρχιτέκτων arkhitekton "architect", from ἀρχι- «chief» and τέκτων «builder», is both the process and the product of planning, designing, and constructing buildings and other physical structures. Architectural works, in the material form of buildings, are often perceived as cultural symbols and works of art. Historical civilizations are often identified with their surviving architectural achievements.

«Architecture» can mean:

- A general term to describe buildings and other physical structures.
- The art and science of designing buildings and (some) nonbuilding structures.
- The style of design and the method of construction of buildings and other physical structures.
- A unifying or coherent form or structure.
- Knowledge of art, science, technology, and humanity.
- The design activity of the architect,[3] from the macro-level (urban design, landscape architecture) to the micro-level (construction details and furniture). The practice of the architect, where architecture means offering or rendering professional services in connection with the design and construction of buildings, or built environments.

Astronomy

From Greek: is a natural science that studies celestial objects and phenomena. It applies mathematics, physics, and chemistry, in an effort to explain the origin of those objects and phenomena and their evolution. Objects of interest include planets, moons, stars, galaxies, and comets; while the phenomena include supernovae explosions, gamma ray bursts, and cosmic microwave background radiation. More generally, all astronomical phenomena that originate outside the Earth's atmosphere are within the purview of astronomy. A related but distinct subject, physical cosmology, is concerned with the study of the Universe as a whole.

Astronomy is the oldest of the natural sciences. The early civilizations in recorded history, such as the Babylonians, Greeks, Indians, Egyptians, Nubians, Iranians, Chinese, and Mayans, performed methodical observations of the night sky. Historically, astronomy has included disciplines as diverse as astrometry, celestial navigation, observational astronomy and the making of calendars, but professional astronomy is now often considered to be synonymous with astrophysics.

During the 20th century, the field of professional astronomy split into observational and theoretical branches. Observational astronomy is focused on acquiring data from observations of astronomical objects, which is then analyzed using the basic principles of physics. Theoretical astronomy is oriented toward the development of computer or analytical models to describe astronomical objects and phenomena. The two fields complement each other, with theoretical astronomy seeking to explain the observational results and observations being used to confirm theoretical results.

Astronomy is one of the few sciences where amateurs can still play an active role, especially in the discovery and observation of transient phenomena. Amateur astronomers have made and contributed to many important astronomical discoveries, such as finding new comets.

Cartography

From Greek khartēs, "papyrus, sheet of paper, map"; and graphein, "write", is the study and practice of making maps. Combining science, aesthetics, and technique, cartography builds on the premise that reality can be modeled in ways that communicate spatial information effectively.

The fundamental problems of traditional cartography are to:

- Set the map's agenda and select traits of the object to be mapped. This is a concern of map editing. Traits may be physical, such as roads or land masses, or may be abstract, such as toponyms or political boundaries.
- Represent the terrain of the mapped object on flat media. This is a concern of map projections.
- Eliminate characteristics of the mapped object that are not relevant to the map's purpose. This is a concern of generalization.
- Reduce the complexity of the characteristics that will be mapped. This is also a concern of generalization.
- Orchestrate the elements of the map to best convey its message to its audience. This is a concern of map design.

Modern cartography constitutes many theoretical and practical foundations of geographic information systems.

Civil engineering

A professional engineering discipline that deals with the design, construction, and maintenance of the physical and naturally built environment, including working on roads, bridges, canals, dams, and buildings. Civil engineering is traditionally broken into a number of sub-disciplines. It is the second-oldest engineering discipline after military engineering, and it is defined to distinguish non-military engineering from military engineering. Civil engineering takes place in the public sector from the municipal level through to national governments, and in the private sector from individual homeowners through to international companies.

REFERENCES

1. Allen, Harold Don. The Teaching of Trigonometry in the United States and Canada: A Consideration of Elementary Course Content and Approach and Factors Influencing Change. Thesis: Rutgers University-Graduate School of Education. Vol. 1, 1977.
2. Joyce, David E. Dave's Short Trig Course. Department of Mathematics and Computer Science, Clark University. Retrieved 3 March 2005.
3. Maor, Eli. Trigonometric Delights. New Jersey: Princeton University Press, 1998.
4. www.clarku.edu/~djoyce/trig/
5. www.themathpage.com/aTrig/trigonometry.htm
6. www.intmath.com/Trigonometric-functions/Trig-functions-intro.php
7. www.mathtv.com
8. www.cadproblems.us/forms/Schwendau_Trig_Chart.pdf
9. www.brightstorm.com/math
10. Trigonometry for Dummies by Mary Jane Sterling.
11. Algebra and Trigonometry: Structure and Method by MCDOUGAL LITTLE.

12. Akkoç, H. & Akbaş Gül, N. (2010). Analysis of a teaching approach aiming at eliminating student difficulties with radian. Ankara University Journal of Faculty of Educational Sciences, 43(1), 97-129.

13. Blackett, N., & Tall, D. (1991). Gender and the versatile learning of trigonometry using computer software. In F. Furinghetti (Ed.), Proceedings of the Fifteenth Conference of the International Group for the Psychology of Mathematics Education (Vol. 1, pp. 144-151).

14. Brown, S. (2005). The trigonometric connection: Students' understanding of sine and cosine. (Unpublished doctoral dissertation). Illinois State University.

15. Choi-koh, S. S. (2003). Effect of a graphing calculator on a 10th-grade student's study of trigonometry, The Journal of Educational Research, 96(6), 359-368.

16. Gür, H. (2009). Trigonometry Learning. New Horizons in Education, 57(1), 67-80.

INDEX